The Radio Today guide to the

Icom IC-9700

By Andrew Barron ZL3DW

Now updated to include firmware v1.30 (D-Star picture sharing mode, Preset mode, customized buttons, one more fixed panadapter range per band, and two new spectrum scroll modes), this book includes useful tips and tricks for the configuration and operation of the fabulous Icom IC-9700 transceiver. Rather than duplicate the manuals which describe each button, function, and control, I have used a more functional approach. This is a "how to do it" book with easy to follow step by step instructions.

The author has no association with Icom, any Icom reseller, or service centers. The book is not authorized or endorsed by Icom or by any authorized Icom Dealer or repair center. Research material for the creation of this document has been sourced from a variety of public domain Internet sites and information published by Icom including the Basic Manual and the Full Manual. The author accepts no responsibility for the accuracy of any information presented herein. It is to the best of my knowledge accurate, but no guarantee is given or implied. Use the information contained in this book at your own risk. Errors and Omissions Excepted.

Radio Today is a trademark of the Radio Society of Great Britain www.rsgb.org and is used here with their kind permission.

Cover graphics by Kevin Williams M6CYB. The transceiver image and Icom logo are used with permission from Icom (UK) Ltd.

Icom, Icom Inc. and the Icom logo are registered trademarks of Icom Incorporated (Japan).

Microsoft and Windows are registered trademarks of Microsoft Corporation.

All other products or brands mentioned in the book are registered trademarks of their respective holders.

The Radio Today guide to the Icom IC-9700

Table of Contents

OTHER BOOKS BY ANDREW BARRON

The Radio Today guide to the Icom IC-705

The Radio Today guide to the Icom IC-7300

The Radio Today guide to the Icom IC-7610

The Radio Today guide to the Yaesu FTDX101

Testing 123
Measuring amateur radio performance on a budget

Software Defined Radio
for Amateur Radio operators and Shortwave Listeners

Amsats and Hamsats
Amateur radio and other small satellites

An introduction to HF Software Defined Radio
(out of print)

ACKNOWLEDGMENTS

Thanks to my wife Carol for her love and support and to my sons James and Alexander for their support and their insight into this modern world. Thanks also to Icom who produced the excellent IC-9700 transceiver and finally, many thanks to you, for buying my book.

ACRONYMS

Amateur radio is chock full of commonly used acronyms and TLAs (three letter abbreviations :-) They can be very confusing and frustrating for newcomers. I have tried to expand out acronyms and explain abbreviations the first time that they are used. Near the end of the book, I have included a comprehensive glossary, explaining the meaning of many terms used in the book. My apologies if I have missed any.

The Icom IC-9700 transceiver

Congratulations on buying or being about to purchase the amazing Icom IC-9700. It is always exciting unboxing and learning how to use a new transceiver. The IC-9700 is a groundbreaking amateur radio transceiver. It is the first full power, multi-mode, VHF/UHF amateur radio transceiver to be based on SDR technology. The receiver uses direct digital sampling for the 2 m and 70 cm bands and a single down conversion stage followed by direct digital sampling for the 23 cm receiver. The transmitter uses direct digital up-conversion for the 2 m and 70 cm bands and a heterodyne up-conversion stage for the 23 cm band. The IC-9700 is also the first VHF and UHF transceiver to feature a panadapter spectrum and waterfall display.

The IC-9700 is in the same family as the Icom IC-7300 sharing an almost identical front panel layout, weight, and size. The release of the IC-7300 created something of a revolution in the amateur radio world, bringing the benefits of SDR technology to thousands of hams. With this radio, Icom provides the perfect companion to the IC-7300. It is a transceiver for VHF and UHF DX operation, working satellites, EME, and repeater operation, supporting both traditional FM repeaters and D-Star digital. Many of the features included in the IC-9700 have never been available on VHF/UHF amateur radio transceivers before.

If you already own an IC-7300 or an IC-7610 you will be familiar with the touch screen display and the positions of many controls. But you will quickly find that the IC-9700 is very much more complicated than either of those radios. This radio is crammed with exciting new features. For a start it has two receivers, so you can receive signals on two bands at the same time. It is also capable of full-duplex crossband operation. Being able to receive signals on one band while transmitting on another band. There is provision for the connection of a 3rd party GPS receiver for use with the D-Star and digital voice gateway modes and for connection of an external 10 MHz frequency reference for superior oscillator stability. The high-power output on the 2 m and 70 cm bands will be great for long-distance work and Meteor Scatter operation. There are an astounding number of memory slots and scanning functions. 99 channels per band, plus 99 satellite memories, 300 GPS memories, and 2500 D-Star repeater memories.

Icom has borrowed some functions from the IC-7300 that you don't normally see in VHF/UHF radios. You get a voice message keyer and recording functions and even an RTTY decoder. Unfortunately, just like the IC-7300, the built-in RTTY is crippled by the lack of any way to connect an external keyboard. You can't easily enter the other station's callsign or chat. I would even settle for the ability to use the onscreen QWERTY keyboard. Perhaps this can be added in a future firmware release.

This revised edition of the Radio Today guide to the Icom IC-9700 includes the new features added in the February 2021 v1.30 firmware update. These include the very useful Preset mode, which can be used for all external digital modes, scrolling the spectrum display if you tune off the edge of the currently displayed span, four fixed edges for the FIX display, the D-Star picture sharing mode, and the ability to apply custom functions to three of the front panel buttons and two microphone buttons.

If this is your first SDR I am sure that you will be excited about the panadapter. I particularly like the 'FIX' spectrum display mode where you can pre-set a band segment. For example, you might set a band edge to cover the SSB part of the 2 m band, a satellite linear transponder downlink, or the segment used by EME enthusiasts.

For CW operators, there is full or semi break-in keying from 6 to 48 wpm and eight message memories including an auto-incrementing contest number. I'm sure that the message memories will be very useful for CW operation through linear satellite transponders. The CW mode functions are lifted directly from the IC-7300 and are much more comprehensive than you would find on other VHF/UHF radios.

Compared to other transceivers there are quite a few changes to the way that you operate the radio. These are mostly due to the touch screen controls. As I have got to know the radio by using it and through delving into every control and menu setting, I have discovered many clever design features. You can certainly see the benefit of decades of Icom technical development and experience.

The radio has a fairly simple front panel layout, with many settings being accessed through the touch screen display. For example, there are no physical 'band change' or 'mode' buttons. Less often changed settings are available through the MAIN, FUNCTION, and QUICK menu buttons. It can be a bit of a challenge remembering which settings are on what sub-menu. The radio has a lot of operating features, so the menu structure is quite large. However, most settings only need to be adjusted once and many can be left at the default settings. Learning about all the neat things the radio can do is what this book is all about. There are step by step instructions for setting up the radio and I have included a 'quick reference guide' at the back so that you can easily find your way through the menus to make the changes that you want.

The IC-9700 is a great radio for satellite operation, EME, chasing DX, contesting, or just chatting on the local repeater. It is a radio for the serious VHF/UHF operator. It is small and light enough for stationary mobile, campervan, portable, Field Day, or maybe even SOTA activation. I enjoy alternating between the fixed spectrum scope which shows you all the signals across a 1 MHz segment of the band and the 'Center' display which shows you a few kHz either side of the VFO center frequency.

The radio has virtually every option that you could want. It combines cutting edge SDR technology with all the controls and features that experienced amateur radio operators expect from an Icom transceiver. Of course, that makes setting it up quite a complicated task. There are many opportunities to configure the radio to your tastes.

The configuration settings are available through the touch screen menus and they are in plain English (or Japanese) rather than the cryptic codes found on transceivers with smaller displays.

I found that I was changing a lot of settings as I customized the radio to my liking. I began to lose track of the default settings, although they are all in the manual. I decided to write down the way that I have the radio configured for future reference. You can decide on your own preferences. I left room for you to add your preferred settings to the relevant tables.

If you use satellite tracking software, D-Star data, or digital mode software for modes such as PSK or FT8, I am sure that you will find the section on configuring the USB CI-V interface between the radio and your PC very helpful. As well as the USB type B interface the radio features an Ethernet LAN connection used for NTP time synchronization, data communication in D-Star mode and DD (digital data) mode, and remote control using the RS-BA1 software. It can also be used to output receiver audio or the 12 kHz I.F signal.

I cover some "weird" operating behaviors in the troubleshooting chapter. If you experience something strange, have a look to see if it is covered in that chapter.

TECHNICAL FEATURES

The IC-9700 is an SDR (software defined radio) or more correctly a direct digital sampling radio. If you haven't used an SDR transceiver before, you will be impressed by how clean the receiver sounds and you will quickly get used to the advantages of the panadapter display. Icom has kept the layout very similar to their previous radios and anyone transitioning from an IC-7300 or an IC-7610 will be very comfortable. One or two controls are available as conventional buttons and also as Soft Key functions. For example, you can turn on the preamplifiers using the P.AMP button or by pressing the FUNCTION button and using the P.AMP Soft Key. Many other controls are only available via the touch screen.

The radio is a 'Tri-bander' with a transmitter capable of transmitting 100 watts on the 2 m band, 75 Watts on the 70 cm band and 10 Watts on the 23 cm band. The display is a 93 x 52 mm color TFT touch screen. It displays a very clear, crisp image, with excellent contrast and color saturation. There is an adjustable LCD backlight to ensure that the brightness can be adjusted to suit indoor or outdoor operation.

Direct digital sampling is used for both the receive and transmit signal paths. The receiver chain starts with a switchable preamplifier, an optional attenuator, and a bandpass filter for each of the three bands. The 2m and 70 cm bands are sampled by the ADC directly.

The 23 cm receiver employs a 929 MHz local oscillator and mixer to shift the received signals down to a 311-371 MHz I.F. which is within the frequency range that can be sampled by the ADC.

The next stage is a pair of RF (radio frequency) switches that enable you to select any two out of the three available bands. The radio can receive signals on two bands at the same time. The output of both switches is sent to a 14-bit LTC2156-14 dual ADC (analog to digital converter) followed by the FPGA (field programmable gate array) which performs digital down conversion. A separate DSP (digital signal processing) chip is used for D-Star digital voice and data, and a CPU (central processing unit) takes care of the control functions and the touch screen display. The audio data stream is converted back into an analog audio signal using an audio frequency DAC (digital to analog converter) and sent to an audio amplifier to drive the speaker and the headphone jack.

The IC-9700 can receive frequencies on two different bands at the same time, but it cannot receive two frequencies on the same band at the same time.

The transmitter uses an audio frequency ADC to convert analog signals from the microphone or other analog audio input into a digital signal. This is followed by digital signal processing and digital up-conversion in the FPGA. After that, the digital signal which is already carrying the transmission at the wanted transmit frequency is converted to an analog RF signal using a very high-speed DAC (digital to analog converter). This is followed by an RF switch which directs the signal to the appropriate RF power amplifier. The 23 cm band transmit signal comes out of the DAC on a frequency between 305.632 MHz and 365.632 MHz. Prior to the driver stage it is upconverted to the wanted frequency on the 23 cm band by using a mixer and fixed oscillator on 1605.632 MHz, before being applied to the input to the 10-Watt 23 cm power amplifier.

Each power amplifier output is fed through a low pass filter appropriate for the band to an individual antenna connector. Each band has its own antenna connector.

The radio supports SSB (USB and LSB), CW, RTTY, AM, FM, DV (digital voice), and DD (digital data to 128 Mb on the 23 cm band only), operating modes.

The IP+ mode is designed to improve the receiver's intermodulation performance by turning on the ADC 'randomization' function. It improves the lab test results, but in real-world conditions, it is usually unnecessary.

Some online commentators say that they don't like touch screens. They find them fiddly and complain of fingerprints. This is NOT a radio for them. It is impossible to use the radio without touching the screen. While one or two touch screen functions are duplicated with "real" buttons, all of the menu structure and even basic functions like changing the operating mode can only be changed via the touch screen display.

Anyway, I really like the touch screen. It works very well. It is crisp, bright, and colorful with very high definition. I find that unless you've been eating jam sandwiches the display does not show fingerprints.

I was impressed with the auto-selection of repeater duplex. If you tune across a repeater output sub-band, the radio automatically selects the correct plus or minus duplex split. This means that if you tune to a repeater output frequency, the radio will transmit on the repeater input frequency. You can save the channel into a memory location without having to manually set the duplex offset. This function can be disabled if you want to transmit in simplex mode on that part of the band. [Auto-selection of repeater duplex may not work, on all bands, in all regions due to variations between country specific band plans].

D-STAR

The Icom Advanced Manual devotes four chapters to D-Star operation. A total of eighty-one pages, compared to the section on Satellite operation which only rates three pages. You can operate D-Star from the normal VFO mode and memory channels, but there is also a special DR (digital repeater) mode which makes accessing and linking to repeaters and reflectors much easier. The DV/DD memory bank can hold 2500 repeater definitions. These can be imported from a .csv spreadsheet file saved to the SD card. Alternatively, you can use the RT Systems IC-9700 programmer software available from RT Systems, or the Icom CS-9700 programming software, which is available as a free download from the Icom (Japan) website. The radio has all the D-Star features found on other D-Star enabled radios. You can work simplex to another D-Star radio, through a local repeater or hotspot, or through a local repeater or hotspot and the Internet to users connected to other repeaters worldwide. The system works very well with no noticeable latency. On the 23 cm band, this type of operation can even be used to provide a low-speed data or Internet connection via your transceiver and a 23 cm repeater. Your local repeater or hotspot has to support D-Star, but Internet-linked repeaters and reflectors don't have to be D-Star specific. Some of them are for DMR, System Fusion, or DMR users. This allows you to talk to hams who are using radios employing a different type of digital voice radio.

You can connect an NMEA compliant GPS receiver to obtain accurate GPS location, speed and direction information for transmission over the DV (digital voice) D-Star mode. If you are not intending to operate mobile, you can enter your location information manually.

The location data is also used to determine the closest repeaters to you, and the distance and bearing between your location and the users of connected D-Star repeaters and reflectors worldwide.

You can also use the radio as a D-Star gateway or hotspot, using the LAN connection and connecting directly to the Internet as a gateway or using a D-Star handheld to talk through the radio to a connected gateway. It is a rather expensive way to achieve this compared to buying a cheap DV dongle, but it is possible.

The v1.20 firmware upgrade added a D-Star 'picture sharing mode' which allows you to send and receive pictures at high speed or at a lower speed while also carrying out a conversation.

SATELLITE MODE

Several other radios have included full duplex cross-band satellite mode operation with forward and reverse tracking of the VFO frequencies. But locking the VFOs together in this fashion can't correct for the difference in the Doppler shift between the 70 cm band (or 23 cm band) and the 2 m band. Most satellite operators prefer to use satellite tracking programs which control the radio through the CI-V interface to automate Doppler correction and provide transponder tracking. The IC-9700 is capable of full-duplex satellite operation with forward or reverse tracking with the added advantage of being able to easily move either the main (downlink) or the sub (uplink) VFO independently and then lock the VFOs together again. This means that for the first time, you really can consider manually managing the Doppler shift using the radio alone. Of course, the option of PC control via the CI-V interface is still available. I successfully configured the radio to work with SatPC32 and others have reported success with MacDopplerPRO. One feature not found on other 'satellite capable' transceivers is that while you are not transmitting you can listen to the uplink frequency and the downlink frequency at the same time.

FRONT PANEL DESCRIPTION

The front panel layout is identical to the IC-7300 although the function of some of the controls is different. There are volume, squelch, and RF gain controls for each receiver and a button for selecting a 'call channel' or the Digital Repeater D-Star mode. There is a separate button for enabling tone squelch. Many basic radio functions and all of the menu commands are accessed via the 4.3-inch color TFT touch screen. The display is particularly clear and bright, retaining readability well under bright lighting. Backlight level is adjustable, and you can select either a black or blue background. Band, mode, filter selection, meter selection, and VFO / memory functions are all selected by touching the appropriate areas on the display. Each Soft Key icon brings up a grid of selectable options.

Press and hold the upper volume control knob, to swap the upper (main) and lower (sub) bands over so that you can transmit on the other band. Pressing the upper volume control knob moves the focus from the lower band to the upper band (the same as touching the upper VFO number on the touch screen). Press and hold the lower volume control knob, to turn the second receiver on or off. Pressing the lower volume control knob moves focus from the upper band to the lower band (the same as touching the lower VFO number on the touch screen).

A row of hardware buttons along the bottom of the display provides access to the main menu, function settings, the spectrum scope, and "quick" selection of a number of parameters. The Exit button can be used to turn off the spectrum and waterfall display or to exit the menu and sub-menu screens.

Pressing the MULTI rotary control allows you to set adjustable functions such as transmit power level and microphone gain. Rotating the MULTI knob either steps through the memory channels or it adjusts the VFO frequency in the pre-determined tuning steps. The choice is selected using the kHz/M-CH button.

The display shows 33 status indicators and functions, and that is without considering the panadapter spectrum scope and waterfall display. If the second receiver is active, both receiver frequencies are shown continuously. Each VFO has a mode and a filter Soft Key, and 'bar meters' to show received signal strength or various selectable transmit functions. Simultaneous metering of multiple transmit functions can also be selected. If a repeater or split offset is in use, the VFO display changes to the transmitter frequency while you are transmitting.

A very comprehensive 'Set' mode allows tailoring of an enormous number of functions. These are all accessed via the touch screen display with nested menu and sub-menu items. The menu structure is in plain language which is so much easier than dealing with the cryptic codes used by earlier transceivers. A keyboard is displayed on the touch screen when alphanumeric data needs to be entered. This can be in either a full QWERTY or a 10-key format which makes data entry very straightforward.

Tuning is very smooth and easy using the 47 mm diameter rotary control knob. It has an adjustable drag. I have not adjusted it as I am happy with the factory setting. The normal tuning rate is either in 10 Hz or 1 Hz steps. You can select fast tuning rates from 100 Hz per step up to 100 kHz per step. The tuning rate speeds up if you rotate the VFO knob quickly. A slower quarter-speed tuning rate can be selected for fine-tuning on CW and data modes.

The microphone connector is the standard 8-pin format and there is a 3.5 mm headphone jack. A full-size SD memory card slot is provided for storing various items such as received and transmitted audio files, voice memory stores, RTTY decode logs, memory contents, and setup data. Screen images can also be captured as PNG or BMP files. The SD card is not provided.

You will have to buy one. SD or SDHC cards up to 32 GB can be used. But capacity does not have to be very large. I have saved files for three different Icom radios on a 16 Gb SDHC card and they only take up 30 Mb in total. So, an 8 Gb SD card will be more than adequate. Make sure that you buy a full-size SD card or one that includes an adapter.

REAR PANEL DESCRIPTION

The rear panel includes a SO-239 antenna socket for the 2 m band and Type-N connectors for the 70 cm and 23 cm bands. There is an SMA connector for the connection of a 10 MHz high stability oscillator. This has caused a high degree of angst and comment on various Internet forums. See my comments about oscillator stability in the troubleshooting section.

An Ethernet connector is provided for connection to a LAN. It can be used for remote control of the transceiver using the RS-BA1 V2 software. The LAN connection is also used if you want to turn the radio into a D-Star gateway or hotspot and for the DD (digital data) mode, and it can be used to connect to a NST time server on the Internet, keeping the clock display very accurate.

The type B USB 2.0 port is used for CI-V computer control and for audio (or 12 kHz I.F.) connections between the radio and your PC. One cable carries all the control and audio data. The radio establishes two virtual COM ports and an audio codec which looks like a soundcard to digital mode and satellite tracking software. The addition of a Type A USB port on the rear panel would have enabled the use of an external keyboard which would be a boon for users of the CW memory keyer and built-in RTTY decoder. These functions are severely limited by the inability to enter a remote station's call sign.

A 3.5 mm stereo phono 'Key' jack is provided for the connection of a CW key or paddle, there is a standard Icom accessory jack, an old-fashioned CI-V 'Remote' jack, speaker jacks for the main and sub receivers (mono!) and a data jack for the connection of an external GPS receiver. This input is for location data, not for a GPS referenced clock.

There is no transverter output or external monitor output on the rear panel. A transverter output would have been handy for anyone wishing to use the radio use as an exciter and converter receiver for the microwave bands. It would have been especially welcomed by satellite operators wanting to use the new QO-100 geostationary satellite. The ability to plug in an external monitor may be missed by some, although it is of very limited appeal on the IC-7610. There are no separate ALC or PTT jacks for connection to a linear amplifier, but these signals are available on the ACC jack.

RECEIVER FEATURES

Both receivers are identical and all receiver functions such as filtering or noise rejection etc. are duplicated and separately adjustable by touching the top (main) or bottom (sub) receiver pane on the display.

Incremental tuning (RIT) is adjusted with the Multi knob and auto-tuning is provided on CW. The same button can turn off the AFC when the transceiver is in FM or DV mode.

There are 99 regular memory channels and two scan edge channels for each of the three bands. Memory access is very straightforward. Each memory channel can be assigned a name up to 16 characters in length and this is quick and easy with the on-screen keyboard.

A quick-access 'memo-pad' stack of 5 or 10 frequencies is also included. Also, there are two Call channels per band for quick access to your most often used frequencies or repeaters.

A host of scanning functions are provided. Filtering functions are very comprehensive. A 'FIL' touch-button on the display scrolls through three pre-set IF filter bandwidths with separate settings for each mode. They can be individually adjusted to suit your requirements. Sharp and soft passband shapes are available at the touch of a Soft Key. Twin PBT allows either side of the filter passband to be shifted independently, shifting or narrowing the overall shape to assist in combating adjacent channel interference. A manual notch filter operates inside the AGC loop preventing desensitization with strong carriers. It has excellent depth with wide, medium, or narrow width settings. A separate auto-tuning notch filter operating in the audio passband removes multiple tones effectively.

An adjustable noise reduction system reduces background noise and improves readability in certain situations. A separate noise blanker eliminates pulse-type noise from car ignition and other sources. Three separate AGC time constants are selectable from a menu of 13 different values (0.1 to 6 seconds). They can be set separately for all modes except FM. The AGC can also be switched off if required.

The receiver audio response can be tailored independently for each mode. The high-pass and low-pass roll-offs can be adjusted separately and the bass and treble responses cut or enhanced.

TRANSMITTER FEATURES

Transmitting functions for SSB include a speech compressor, VOX, and a transmission monitor. The audio transmit filter bandwidth may be set to menu-adjustable wide, mid, or narrow settings. In addition, the bass and treble responses can be cut or enhanced separately for each voice mode in a similar fashion to the receive audio. The voice keyer stores eight pre-recorded audio messages and operates in any of the voice modes including the voice-data modes. Each message has 90 seconds of recording time. Operation and recording are the same as on the IC-7300.

On CW there is the usual provision for full and semi break-in with adjustable drop back delay. The keying envelope rise and fall times are adjustable between 2 and 8 ms and an additional 'transmit delay' is selectable for each band to avoid 'hot-switching' linear amplifiers or other accessories. A CW message keyer is included operating over the speed range 6 – 48 wpm with adjustable weighting and a variety of keying paddle arrangements. Eight memories store up to 70 characters each with provision to send an automatically incrementing serial number and auto-repeat after a time delay. The RTTY mode includes a built-in RTTY decoder and an eight message keyer similar to the CW one. The message keyers are identical in configuration and operation to the ones in the IC-7300.

FM operation includes automatic duplex selection [*in some regions*], CTCSS tone repeater access, tone squelch, and repeater split operation.

For digital mode operation using external software, there are three data modes. The modulation input can be taken from the USB cable, microphone socket, or accessory socket. The USB-D mode is used for AFSK modes such as FT8, JT65, and PSK. The FM-D mode would be used for Packet and APRS modes.

Approach

Rather than duplicate the manuals which describe each button and control, I have used a more functional approach. This is a "how-to" book. For example, I describe how to set up the transceiver for SSB operation, including all the relevant menu settings. Then I follow that up with setting up instructions for CW, FM, RTTY, DV, DD, and external digital mode software such as JT65 or FT8.

Although the IC-9700 has a simple front panel and is easy to operate, the touch screen controls will be new to many and there are a lot of configuration options. Especially if you want to use external satellite tracking or digital mode software such as WSJT, WSJT-X, Fldigi, MixW, CW Skimmer, or MRP40 to mention a few. I was quite surprised at the number of things that had to be configured. Icom has allowed you to make things exactly the way you want them. I imagine that you will experiment with some of the settings more than once before you decide on the optimum settings for your radio.

There are sections on updating the radio firmware, loading the Windows driver software for the USB cable connection, connecting the radio to your PC for CI-V control, and setting the clock. I cover satellite operation, the DV mode, and the FM repeater mode. Including how to set up tones for repeater access and store repeater channel frequencies in the radio memory slots.

The 'Setting up the radio' chapter is followed by information about the front panel and touch screen controls and the MULTI menus. Then a section about 'Operating the radio' in various modes including Split operation. After that comes information about every MENU item and FUNCTION setting.

The 'Useful Tips' section describes the screen saver and a special setting for the spectrum display.

The Troubleshooting section deals with some oddities that might trip you up.

The Glossary explains the meaning of the acronyms and abbreviations used throughout the book and the Index is a great way of going directly to the information that you are looking for.

Last, of all, there is the Quick reference guide, where you can look up the menu steps for most of the commonly used functions. I find it useful because I can never remember where to find the settings I want.

Conventions

The following conventions are used throughout the book.

Front panel controls and buttons are indicated with a highlight. TRANSMIT

Touch screen controls are indicated in uppercase without a highlight. AGC MID

'Touch' means to briefly and gently touch the item on the touch screen. There is no need to press hard on the screen.

'Touch and hold,' means to keep touching the item on the touch screen for one second or until the function changes. It usually opens a control option or sub-menu window on the touch screen.

'Press' means to press a physical button or control. 'Press and hold' or 'hold down' means to hold a physical button or control down for one second or until the function changes. It usually opens a control option or sub-menu window on the touch screen.

If 'Beep' is set to on, all touch and press operations are accompanied by a quiet beep. Touch and hold or press and hold operations are often accompanied by two beeps.

The main MENU and the FUNCTION menu both have two levels <1> and <2>. If the menu selection you are looking for is not visible, select the other option.

Words in <brackets> indicate a step in a sequence of commands. The sequence usually begins with pressing a <button> or a <knob> followed by touching one or more <Soft Keys> or <icons> on the touch screen. A 'Soft Key' is an icon on the touch screen that represents a button or control.

The 'Return' or 'Exit' Soft Key is a back-turning arrow. Something like this ↰.

'Mode' usually means one of the transceiver modulation modes. SSB, CW, RTTY, AM, FM, DV, DD, or data modes (USB-D, LSB-D, AM-D, or FM-D). But it can also mean the 'CENT' band scope mode, 'FIX' panadapter mode or an external digital mode such as JT65 or PSK performed by software running on your PC.

'Panadapter' or 'Spectrum Scope' means the spectrum and waterfall display. The radio can display either a 'Band Scope' in which the main VFO is always in the center and the scope displays the spectrum below and above the VFO frequency. Or a 'Panadapter' in which the receiver can be tuned to a frequency anywhere across the range of frequencies that are displayed. To avoid confusion, I use the term 'panadapter' whether the display is in 'CENT' band scope mode, or in 'FIX' panadapter mode.

Menu screens

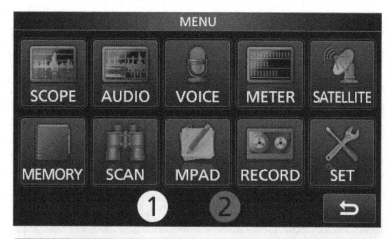

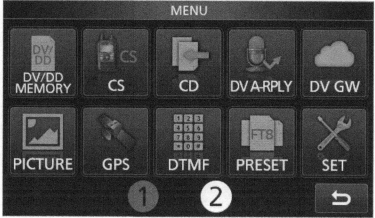

Main MENU screens (FM mode)

Each menu function is discussed in the MENU chapter (page 97).

Some of the Soft Key icons change depending on the radio mode. In CW the VOICE icon becomes KEYER, and in RTTY mode it becomes DECODE.

The page two icons are mostly for D-Star operation. The DTMF icon is only enabled when the radio is in FM or DV mode.

Function screens

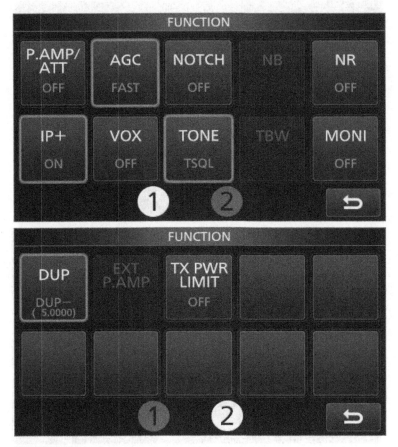

FUNCTION menu screens

Each function is discussed individually in the FUNCTION chapter (page 141).

When a function is active it is shown ringed in blue or, in the case of the front-end attenuator, in amber. Some functions are disabled in some modes. For example, the FM mode does not support the manual notch filter, the noise blanker, or variable transmit bandwidth. SSB does not use the duplex (DUP) setting. In this example, the external pre-amplifier power feed is disabled by a setting in the main menu.

Setting up and using the radio

The Icom manuals do a good job of identifying all the controls and menu items, but they lack 'step by step' instructions on how to set each control correctly. In this chapter, I cover the processes that you need to follow to get ready for operating the radio.

Each section lists the strings of menu commands that you need to follow when you set the audio levels, tone settings, and transmitter bandwidth. It is important to set the Mic Gain and Compression controls correctly for SSB operation. I also explain the setup for the FM, CW, RTTY, DV and DD modes including the built-in options and using external digital mode programs. There are instructions on configuring the eight voice messages for SSB and the pre-defined keying messages for CW and RTTY.

I tell you how to download and install the USB driver software to allow the radio to communicate with your computer and discuss setting up the virtual COM ports and the Audio CODEC, linear amplifier connections, and the increasingly popular FT8 mode. Finally, I cover formatting and using the SD card, how to do firmware updates, and how to set the clock.

The radio is very easy to configure once you know which menu settings to use. Where I have changed my radio from the Icom default settings, I explain why I made the change and the effect that it has on the radio. The big advantage of using the instructions in this section rather than using the Icom manual is that I have included all the necessary steps, and the optional ones, in one place. I tell you what the controls and settings do and how they should be adjusted.

DISPLAY YOUR CALLSIGN WHEN THE RADIO STARTS

It is nice to personalize the radio by having it display your name or callsign as the radio 'boots up.' You can choose what is displayed upon start-up.

- Leave <MENU> <SET> <Display> <Opening Message> set to <ON> if you want to see the Icom IC-9700 screen on start-up. "Of course, you do!"

- Setting <Power ON Check> to <ON> displays the current RF Power setting.

- If you have added your callsign in the 'My Station' menu, it will be displayed at the bottom of the screen during startup. Select <MENU> <SET> <My Station> <My Call Sign (DV)>. Touch and hold the top line and select <Edit> or <Add>. Type in your callsign /9700, then press <ENT>. If you won't be using D-Star, you don't need to add the /9700 suffix.

SETTING UP THE RADIO FOR SSB OPERATION

There are two settings that should be adjusted before transmitting on SSB. You need to set the Mic Gain control correctly and the Compression level control. You don't have to use the compressor if you are primarily interested in Rag Chewing, but you might as well set it anyway. The instructions in the manuals are a bit vague, but the process is simple, and you only need to do it once. Like the IC-7610 and unlike the IC-7300, the default levels turned out to be much too high for my voice using the supplied hand microphone. The good news is that the radio software will not allow you to exceed 100% output power and the ALC (automatic level control) limits the audio modulation level so that it will not distort the modulation. So even if you leave the Mic Gain at the default (50%) setting, the radio will probably sound OK. However, leaving the Mic Gain at 50% may result in low average power output. See the note about low average RF power in the troubleshooting section, on page 203.

In the absence of a clear guide, I have devised my own method for setting the levels. This may not be how Icom or other commentators do it, but it works for me.

1. First, we will set the MIC GAIN control on SSB mode with the Compressor turned off. You can do this adjustment on any band.

    a. With the radio connected to a 50 Ohm dummy load or a tuned antenna. Set the radio for the correct mode USB (upper sideband) on a frequency in the 'SSB' part of the band, outside the repeater input segment. Make sure that you have not selected a data mode. If the blue icon says USB-D touch DATA to set it back to the voice mode.

    b. Press the MULTI knob and touch the RF POWER Soft Key to turn the icon blue. Turn the Multi knob to set the RF POWER to 100%.

    c. If the compressor is turned on, touch the COMP Soft Key on the MULTI sub-menu to turn the compressor off (COMP OFF). Exit the MULTI menu.

    d. I like to see all the metering at once by selecting <MENU> <METER>. But if you like you can just select the relevant meter reading by touching the meter scale. Repeatedly touching the meter scale cycles through Po (power out), SWR, ALC, COMP, V_D, and I_D. Press M.SCOPE to get rid of the panadapter display.

    e. Press MULTI to show the MULTI menu and touch the MIC GAIN Soft Key to activate the control.

    f. Hold the microphone and press the PTT button. Speak into the microphone in the same way that you would while making a contact or calling CQ. Watch the ALC reading on the Multi-Function meter or if you are not using the multi-function meter select ALC on the meter.

While speaking, adjust the MIC GAIN while observing the ALC reading. The ALC meter reading should traverse the whole of the red marked zone occasionally peaking at or near to the top of the zone. It should **not** be compressed up to the top end of the range. Running the ALC at a high level will not give you more output power. The RF Power output should occasionally peak to 100%.

If the RF power never reaches 100% the microphone level is too low. If the blue ALC metering is always up in the top half of the red line range, the microphone level is too high.

It is good practice to use your callsign and that tell the world that you are "testing," especially if you are transmitting into your antenna.

g. For my station at home, I ended up reducing the MIC GAIN to 24%. Your setting may be higher if you speak more quietly than I do.

2. Next, with the compressor turned on, we will set the Compressor level.

a. If necessary, press MULTI to show the MULTI menu. Then touch the COMP button to turn the compressor ON. A blue indicator beside the Soft Key indicates that compression is on. The text should now say COMP ON. You will also get a COMP indicator above the VFO kHz display.

b. I like to see all the metering at once by selecting <MENU> <METER>. But you can just select the relevant meter reading by touching the meter scale. You will have to check the ALC and COMP readings as you speak.

c. Hold the microphone and press the PTT button. Speak into the microphone in the same way that did while setting the MIC GAIN. It is important that you don't shout. You should speak as you would while on the air. Set the COMP LEVEL control so that the COMP meter ranges in the middle of the zone between 5 and 15 but never peaking over 20. I prefer my voice to sound less compressed, so I set the COMP LEVEL to the default setting of 5 dB which makes the compression a bit lower. On the multi-function meter, I get regular compression up to a reading of 5 with occasional peaks up to 10. The meter does not indicate all the time, only when the compressor is acting on a voice peak. I found the meter to be not very useful, so I just selected 5 dB of compression.

d. With the compressor turned on you should see the transmitter hit 100% more often than it did with the compressor turned off. The ALC meter should be livelier, traversing the entire range marked by the red line and peaking to the top of that range.

If the ALC is constantly hammering the top of the red line scale. Reduce the microphone gain. As long as your voice peaks are hitting 100% RF power you have enough microphone gain. That's all you need to do for SSB transmission.

3.   One really neat feature of the IC-9700 is the voice message keyer macros. If you plan to use the voice message keyer, you will need to set the output level of that so that it modulates the transmitter to the same level as the microphone. You want listeners to be unable to detect when they are listening to a recorded message. Setting up the voice keyer is covered on page 103.

SETTING THE TRANSMIT BANDWIDTH (TBW)

If you are in SSB mode, you can select from a choice of three transmit audio bandwidths. Press the FUNCTION button and then touch the TBW Soft Key to cycle through NAR, MID, or WIDE.

WIDE is suitable for Rag Chewing on the local Net or chatting to locals on the 80m or 10m band. MID is better suited to working DX and the NAR (narrow) mode is suited to contest operation.

If like me, you are happy to use the default settings, that is all you need to know. However, you can set the actual transmit bandwidth for each of the three options if you want to.

➢   SSB tone controls and transmitter bandwidth

-   Select <MENU> <SET> <Tone Control/TBW> <TX> <SSB>

-   You can adjust the Bass and Treble of your transmitted audio. I don't recommend making any big changes, but I looked at the transmit audio spectrum while transmitting and adjusted the treble to +2 to make my transmitted audio spectrum a little flatter. <MENU> <AUDIO> opens the audio scope screen.

-   The next three menu items are for adjusting the transmit bandwidth on Wide, Mid and Narrow.

    o   Wide default is 100 Hz to 2900 Hz

    o   Mid default is 300 Hz to 2700 Hz. (I changed mine to 200 – 2700 Hz)

    o   Nar default is 500 Hz to 2500 Hz. (I changed mine to 300 – 2500 Hz)

-   Touch and hold any TBW setting to reset the bandwidth to the Icom default.

➢   SSB-D data mode transmit bandwidth

-   Select <MENU> <SET> <Tone Control/TBW> <TX> <SSB-D>

The default transmitter bandwidth for the SSB data modes is 2.4 kHz (300 Hz to 2700 Hz). This is less than the 3 kHz (widest) receive filter for the SSB-D data mode. Which means that your external digital mode program will display a wider panadapter than you can use for transmitting. If this bothers you, you could reduce the FIL1 bandwidth of the USB-D1 receiver filter to 2.4 kHz, then they will match. Generally, you will use the middle or narrow receive filter for external digital modes in which case it is not an issue. I decided to leave the TBW at default and to reduce the USB-D1 filter to 2.8 kHz. Halfway between the original 3 kHz receiver filter and the 2.4 kHz transmit bandwidth.

The maximum upper frequency you can set is 2900 Hz and the minimum lower frequency is 100 Hz (2.8 kHz bandwidth). Touch and hold the TBW setting to reset the bandwidth to the Icom default.

➢ AM, FM and DV modes transmit bandwidth

You can't adjust the transmit bandwidth for the AM, FM or DV modes, but you can adjust the transmitted bass and treble.

• Select <MENU> <SET> <Tone Control/TBW> <TX> (<AM> <FM> or <DV>)

SETTING THE RECEIVER TONE CONTROLS

➢ RX HPF/LPF adjustment

You can set audio high pass and low pass filters for all modes except the DATA modes. Select <MENU> <SET> <Tone Control/TBW> <RX> *<choose mode>* <RX HPF/LPF>. I have not made any changes as the default settings seem OK to me. Note that adjusting the RX HPF/LPF will override the bass and treble setting and return those controls to zero.

➢ Bass and Treble adjustment

You can adjust the Treble and Bass for the four voice modes, SSB, AM, FM and DV. Select <MENU> <SET> <Tone Control/TBW> <RX> *<choose mode>* <RX Bass> or <RX Treble>. Note that if you change the bass or treble setting, it will override the RX HPF/LPF setting and return it to default '- - - - - - -.'

MICROPHONE GAIN FOR AM, FM OR DV OPERATION

If you set the microphone gain for SSB operation as suggested above, it should be fine for AM, FM, and D-Star operation. However, if you get complaints that your FM transmission sounds a little weak, you can increase the microphone gain up to 50% with no ill effects. Note that there is only one microphone gain control. It affects all voice modes on all bands.

SETTING UP THE RADIO FOR CW OPERATION

There is no real setup required for the CW mode. No levels to set other than the transmitter power and the sidetone level. Note that the MONI (transmit monitor) function is disabled in CW mode because the sidetone is always turned on. If you don't want sidetone you can turn down the sidetone level to zero. (See note below).

➢ Keys

There are provisions for different types of 'Morse' key. The non-standard 3.5mm rear panel 'Key' jack can be used with a Paddle, Bug, or Straight Key. You can even send CW using the up and down buttons on the Icom hand microphone. The wiring is standard with 'dots' on the phono plug tip, 'dashes' on the ring and common on the sleeve. See page 156 for more information.

➢ CW settings

The CW controls are only accessible when the radio is in the CW mode. The most often used controls for CW operation are the key speed and pitch. Press the MULTI button and select KEY SPEED or CW PITCH. Turning the Multi Knob adjusts the selected item. The CW speed is adjustable from 6 wpm to 48 wpm. The CW pitch is adjustable from 300 Hz to 900 Hz. I use 700 Hz.

The other menu settings for CW are on the KEYER menu. Set the radio to CW mode, then <MENU> <KEYER> <EDIT/SET> <CW-KEY SET>. There are eight menu options spread over two screens. You probably won't need to change any of the options. There is a table of the settings on page 108.

➢ Break-in setting

Your CW (Morse Code) signal won't automatically be transmitted unless either full break-in or semi break-in has been selected. When the radio is in CW mode, press the FUNCTION button and choose either BKIN or F-BKIN and the radio will transmit when you operate the key or trigger a message macro.

I suggest that you don't use full break-in on the 2m band. The noise from the PTT relay will drive you nuts. On the 70 cm band, the keying is done with silent Pin Diode switches and on the 23 cm band, the micro-relay used for PTT is very quiet.

If you want to practice your CW without transmitting, you can listen to the sidetone when break-in is set to OFF. To transmit in that mode, you have to manually key the transmitter by pressing the TRANSMIT button, pressing the PTT button on the microphone, using a CI-V command, or by grounding the SEND line on the ACC jack.

The break-in settings affect the sending of keying macro messages as well as CW sent from a key or paddle. But they have no effect on CW sent from a PC application.

- With BK-IN OFF you can practice CW by listening to the side-tone without transmitting.

- F-BKIN. The full break-in mode will key the transmitter while CW is being sent and will return to receive as soon as the key is released. This 'QSK' mode allows for reception of a signal between CW characters.

- BKIN. The semi break-in mode will key the transmitter while CW is being sent and will return to receive after a delay. Touch and hold the F-BKIN or BKIN Soft Key to adjust the delay. Turn the Multi knob to change the setting. The default is a period of 7.5 dits at the selected CW keying speed.

➢ Sidetone

CW sidetone is 'always on' but you can set the level to zero if it is annoying, or if your key generates its own sidetone. In CW mode select <MENU> <KEYER> <EDIT/SET> <CW-KEY SET> <Side Tone Level>.

➢ CW message keyer

When in CW mode press MENU then KEYER to show the eight CW messages. They can be used for DX or Contest operation or just to save you from having to send the same message over and over. They are great for sending CQ on a quiet band and for establishing a path over a linear transponder satellite.

Touching one of the **M1 to M8 Soft Keys** sends the CW message. Touch again or send a dit or dah from the paddle to stop sending the message.

Touch and hold one of the **M1 to M8 Soft Keys** to keep sending the message until you stop it by touching the Soft Key again or by sending with the key or paddle.

At a minimum, you will have to add your callsign to the keying macros or they will send "Icom" in place of it.

To edit the messages. In CW mode select <MENU> <KEYER> <EDIT/SET> <EDIT>. Touch and hold the message that you want to edit. Then select the <EDIT> option.

An on-screen keyboard will appear, and you can edit or replace the message. Make sure that you press ENT, or your changes will be lost when you exit the keyboard screen.

➢   Contest number CW mode

In the M2 message, you can enter a star (*) after the signal report and the radio will update the contest number used for contest reports automatically. This auto numbering star can only be used in one of the eight macros and by default it is included in macro M2. If the star has been used in any other macro, it is not shown on the list of available symbols when you edit any of the others. The upward arrow under the M2 text indicates that this macro is the one with auto-numbering. If you want to shift the function to another macro, you have to remove the star from M2 first. You also have to change the 'Count Up Trigger' setting to the new macro. It's easier just to leave it on the M2 macro.

If you don't work contests you can change the message to a message you can use and leave out the star.

A Caret ^ symbol removes the space between two letters, for example, ^AR or ^BK.

On the <MENU> <KEYER> <EDIT/SET> <001 SET> menu you can set

a)   The number style used for automatic contest numbers.

   a.   Normal numbers 001 etc.

   b.   ANO style A=1, N=9, O=0

   c.   ANT style A=1, N=9, T=0

   d.   NO style N=9, O=0

   e.   NT style N=9, T=0

b)   The macro using the automatic 'Count Up' feature? The default is M2. It must be the macro that has the star (*) in the text.

c)   The start number, usually 001. Note that if you get a busted contest QSO you can decrement the counter by touching -1 on the Keyer message screen.

Touch the Return icon ↺ or the EXIT button to exit each menu layer.

➢   ¼ tuning speed

The digital modes and CW allow the use of the ¼ tuning function. Select <FUNCTION> <1/4> to turn the function on or off. The function slows down the tuning rate of the VFO to make tuning in narrow CW and digital mode signals easier. It is indicated with a ¼ icon to the right of the 10 Hz digit of the frequency display. It is indicated above the 1 Hz digit if the 1 Hz tuning step is active. Touch and hold the 10 Hz digit on the VFO display to switch to the 1 Hz tuning step.

COMPUTER SOFTWARE

The IC-9700 can communicate with most PC based digital mode and satellite tracking programs via the USB cable and you can use the RS-BA1 v2 remote control program to control it remotely via an Internet connection. There is a server built-in, so a PC connection is not required at the radio end. Icom has released a free programmer called CS-9700 which allows you to easily edit and add to the thousands of memory slots available in the radio, including the scan edges and calling channels. You can also edit the satellite memories and the GPS memory bank. The Icom manual states that the radio comes with a list of worldwide D-Star repeaters pre-loaded, but mine was supplied blank. The radio can store 2500 DV repeaters divided into 50 groups. You can also set most if not all of the menu settings and the text for the RTTY and CW message keyers. The RT Systems IC-9700 programmer has a greater emphasis on memory management, and if you include the column headers you can cut-n-paste memory data directly from Excel. Both programs allow you to import channel data from .csv files.

SETTING UP A CONNECTION TO YOUR PC

The USB port on the rear panel of the radio is used for CI-V CAT control of the transceiver and for transferring audio to and from a PC for external digital mode software. It is a USB 2.0 Type B port. You need a Type A to Type B USB 2.0 cable to connect the radio to a PC. If you have a spare USB 2.0 port on the PC, feel free to use it as there is no advantage in wasting a USB 3.0 port.

➢   Driver software

To use the USB port, it is essential that you download and install the Icom driver software on to your PC. If you have already installed the USB driver for an IC-7300 or one of the other radios on the list. You do not have to load it again. The Icom driver software can be downloaded from the Icom website.
http://www.icom.co.jp/world/support/download/firm/index.html.

Make sure that you download the version that is suitable for the IC-9700. The USB driver is not in the IC-9700 firmware section, it is further down the list. The current version is USB Driver (Version 1.30).

IC-7100/ IC-7200/ IC-7300/ IC-7410/ IC-7600/ IC-7610/ IC-7850/ IC-7851/ IC-9100/ IC-9700	USB Driver(Version 1.30), Driver Utility and manuals.	2018/06/07

The driver software creates two virtual COM ports. On my PC they are COM9 and COM10. The port numbers that the driver software creates may be different depending on what is already in use on your computer. You will need to know the COM port numbers when you set up the digital mode or other PC software. The COM port is used for CI-V (CAT) control of the radio from the PC. Its RTS and DTR lines are used to key the radio to transmit
and to send CW or digital mode data.

Devices

Bluetooth, printers, mouse

In Windows 10 select 'Settings' and then 'Devices.'

On the 'Bluetooth and other devices screen' under the 'Other devices' heading, you should see some new COM ports with rather catchy names. Take a note of the COM port numbers, especially the lowest com port number as it will be the most used.

Silicon Labs CP210x USB to UART Bridge (COM9)

Silicon Labs CP210x USB to UART Bridge (COM10)

If the COM port is not there, the PC is not seeing the radio. Check that the USB cable from the PC is plugged into the USB port on the back of the radio. Try unplugging the USB cable from the PC end and then plugging it back in again. If it still won't work, reload the Icom driver software with the radio plugged in but turned off.

➢ The Audio Codec – Windows 10

In addition to creating the two USB ports, the driver software creates an audio CODEC (coder-decoder). This makes the radio look like an audio device (microphone or speakers) to the computer.

In Windows 10 select 'Settings' and then 'Devices.'

Audio

Speakers (Realtek High Definition Audio)

Devices

Bluetooth, printers, mouse

USB Audio CODEC

You should see the Icom audio codec listed as 'USB Audio CODEC.' If it is not there, the sound won't work. Try unplugging the USB cable from the PC end and then plugging it back in again. If it still won't work, reload the Icom driver software with the radio plugged in but turned off. It worked for me.

Select 'Sound Settings' on the right side of the Windows 10 screen. On the next screen select 'Sound Control Panel.' You should see a device listed as a USB Audio CODEC or something similar. I found the labeling confusing, so I changed the names.

- On the 'Playback' tab you will see the Icom CODEC listed as a USB CODEC. Right-click the icon and select properties. You will be able to change the name and, if you like, choose a different icon. I picked one that looks like a radio with a blue dial. The Playback tab is for sound out of the PC and into the radio. I changed the **Playback** device to '**Icom IC-9700 Input.**'

- Select the Advanced tab and use the drop-down control to select '16 bit, 48000 Hz (DVD Quality)' and then click OK.

- On the 'Recording' tab you will see the Icom CODEC listed as a USB CODEC. Right-click the icon and select properties. You will be able to change the name and, if you like, choose a different icon. I picked one that looks like a radio with a blue dial. The Recording tab is for sound out of the radio and into the PC. I changed the **Recording** device to '**Icom IC-9700 Output.**'

- Select the Advanced tab and use the drop-down control to select '16 bit, 48000 Hz (DVD Quality)' and then click OK.

➢   Radio and COM Port device setting

This is very important and can be a limitation on whether your digital mode or other PC software can "talk" to your IC-9700. Icom uses an address of A2h (Hexadecimal) for the IC-9700.

Some earlier Icom radios used 88h and most radios don't use an address at all. The IC-7610 uses an address of 98h and the IC-7300 uses 94h. You can change the address in the radio using the CI-V settings, but this would be a last resort because it could cause software that is expecting to use A2h to fail.

What all this means is;

- If your digital mode software includes a device called Icom IC-9700. Select that option.

- A setting of IC-7600, IC-7610, IC-7800, or IC-9100 may not work because of the A2h address issue.

- For MixW. Under <Hardware> <CAT Settings>, select CAT = 'ICOM' and Model = 'Other.' Set the 'Addr' to A2h then push OK. The address seems to change randomly after that, but it does not seem to matter. However, if you change any of the settings on the tab, make sure that the address is set back to A2h before you close the tab.

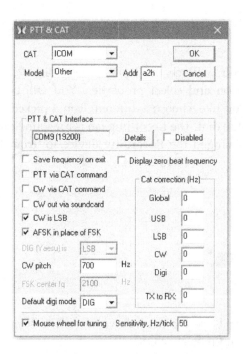

- I set the Com port to COM9, 19200, 8, N, 1, RTS = PTT, DTR = CW.

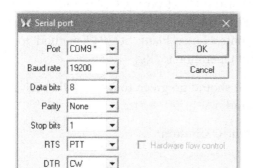

- For MMTTY click the 'Option(O)' tab. Select the 'Setup MMTTY(O). Then select the 'TX' tab and then the 'Radio Command' button on the right side. Set 'Port' to COM9, 'Baud' to 19200, Char wait to 0, Data length to 8 bits, 1 stop, No parity, and uncheck Xon/Xoff, check the DTR/RTS PTT box. Ignore the Commands box.

- Set xx to A2 and Model to Icom CI-V. Group is set to Icom xx=addr 01-7F.

- Save the settings using the Save button, then exit using the OK button.

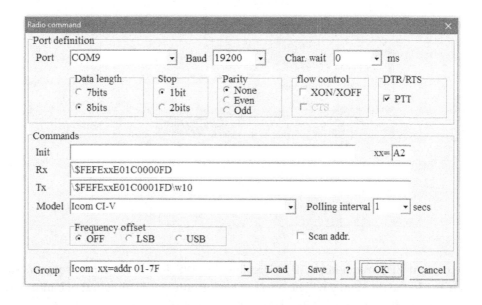

- WSJT-X v2.1.0 supports the IC-9700. Select <File> <Settings> <Radio> and choose the Icom IC-9700. Set the COM port (COM9) and the Baud Rate (19200). Leave Data Bits, Stop Bits, and Handshake set to 'Default.' It is important to leave 'Force Control Lines' blank. Set 'PTT method' to RTS. Set 'Mode' to Data/Pkt. Set Split Operation to 'Rig.'

Press the test CAT button. It should go green to indicate that the software can communicate with the radio.

Exit the setup screen using the OK button.

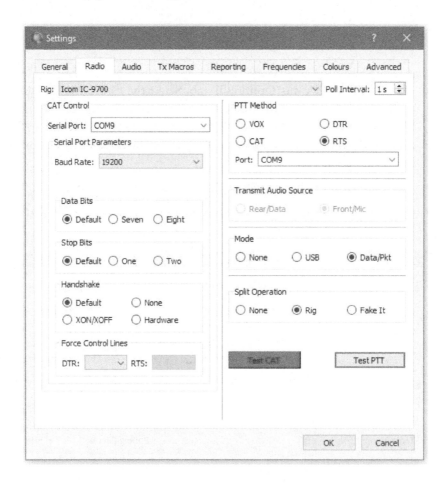

The current VFO frequency will be displayed in WSJT-X. Note that in some modes such as FT8 the frequency will be on a red background indicating 'out of band.' That is because there is no default 2m band frequency for FT8.

➢ MRP40 CW program

For MRP40, select Options, TX settings, Edit COM port pin configuration. COM9, Send pin = DTR, PTT pin = RTS, Activate PTT pin while sending via soundcard checked. Disable PTT Function is unchecked.

➢ N1MM Logger+

I am using N1MM Logger+ V 1.0.7870. It has support for the IC-9700. To set up communications with the radio, Start N1MM. Under 'Config - Configure Ports, Mode Control, Audio, Other - Hardware.' Set the Port dropdown list to the COM port (COM9). Set the Radio dropdown to IC-9700. Check CW/Other.

Click SET. I have speed: 19200, Parity: N, Bits: 8, Stop: 1, DTR: CW, RTS: PTT, Icom code A2, Radio: 1, Delay: 30ms.

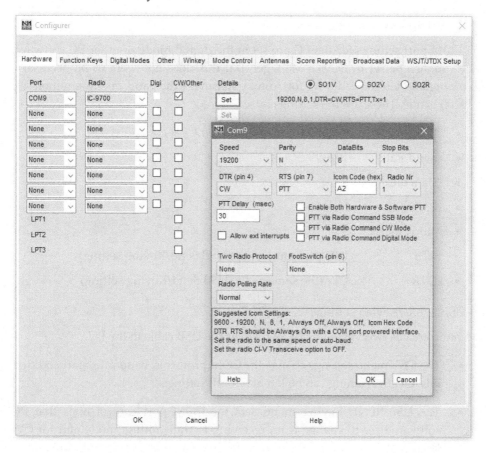

Note: if you want to use the N1MM voice keyer macros instead of, or as well as, the ones in the radio, you will have to set <MENU> <SET> <MOD INPUT> <Data Off Mod> to <MIC, USB>. There are issues with this choice, but it is the only way to get the audio into the radio when it is in SSB or FM mode rather than the corresponding data mode. The main problem is that the microphone is live while messages are being sent. Also, there is a risk that you might forget to use the SSB-D data mode for external digital mode programs and that can cause modulation quality problems.

➢   COM Port settings in the IC-9700 / CI-V settings

The COM port settings in the radio are set in the CI-V menu. After some experimentation, I found that the default settings work well. I recommend leaving everything set to default. There is a full table of CI-V settings in the chapter on 'Special SET menu items' on page 138.

➢   COM Port settings in PC software

The COM port settings for the PC are set in the digital mode or other PC software. They can also be set in Device Manager, but this does not seem to be necessary. Assuming you left the radio set for Auto, you can set the baud rate to anything up to 19200 bauds. These are the settings I use.

- Port   COM 9 (or the nominated CI-V com port)
- Baud Rate   19200
- Data bits   8
- Parity   None
- Stop bits   1
- RTS   PTT (the same as the USB Send/Keying setting)
- DTR   CW (the same as the USB Send/Keying setting)

➢   USB SEND/Keying settings

The radio creates two virtual COM ports over the USB cable to the PC.

- The CI-V menu settings change the COM protocols, baud rate, start and stop bits, etc. Everything can be left at default settings.

- The USB SEND/Keying menu sets the COM port control lines that PC software uses to change from receiving to transmitting mode and for CW keying. It also sets the RTTY keying line if you are using FSK (frequency shift keying using a digital keying signal) rather than AFSK (frequency shift keying using audio tones).

Select <MENU> <SET> <Connectors> <USB SEND/Keying>

- USB SEND is the PTT transmit line. I set it to 'USB (A) RTS.'

- USB Keying (CW) is the line used to send the CW characters once the send line has switched the transceiver to transmit. I set it to USB (A) 'DTR.'

- USB Keying (RTTY) is the line used to send RTTY FSK characters once the send line has switched the transceiver to transmit. I set it to 'USB (B) DTR.'

- Inhibit Timer at USB Connection. This is used to stop the radio sending a SEND, transmit keying signal when the USB cable is first plugged in.

*The RTS and DTR labels don't matter. You can use either line for the transmit PTT as long as you use the other line for CW. The names relate to old-fashioned RS-232 communications between 'old school' computers. RTS stands for 'ready to send' and DTR stands for 'data terminal ready.' But the lines have not been used for that sort of signaling since the 1970s. I always use 'ready to **send'** for the '**Send'** command and '**data** terminal ready' for the CW **data** signal.*

➢ **IC-9700 audio settings -** USB Receiver Output Level

I have chosen to leave the audio level being sent to the PC at the default 50% setting and will adjust the audio levels on the PC if required. If you do want to change the audio level being sent to the PC, select

<MENU> <SET> <Connectors> <USB AF/IF Output> <Output Select> <ON>.

<MENU> <SET> <Connectors> <USB AF/IF Output> <USB AF Output level>.

➢ **IC-9700 audio settings -** USB MOD Level

To avoid the possibility of overdriving the transceiver on digital modes, I have chosen to reduce the audio level being used to modulate the transmitter so that the RF output power is just peaking to 100 Watts. The setting should only have to be done once although you may have to go through the procedure several times to get it right. After the radio is set to your satisfaction you can adjust the PC soundcard output level or the transmit level control on the digital mode software. To make all of you digital mode software output audio at the correct level. That way you can swap programs without having to worry about MOD (modulation) levels.

Unfortunately, you can't display the USB MOD Level menu setting and the transmitter metering at the same time, so you have to set the modulation level, then transmit a digital mode while observing the metering. Then go back to the menu to make an adjustment to the MOD level and then back to the meter display while you transmit again. Eventually, after a few cycles, you will end up with the level set perfectly.

- Before you begin, set the transmitter for 100% power using the MULTI control.

- To set the audio level being sent to the transmitter modulator, select <MENU> <SET> <Connectors> <MOD input> <USB MOD Level>. Initially set it for about 35%.

- Touch and hold the meter scale or press <MENU> <METER> to show the multi-function meter. Check the RF Power output meter and the ALC meter while transmitting a few seconds of your favorite digital mode. I used PSK31.

- The level is set correctly when the transmit RF power indicted on the Po meter is flickering between the red +60 dB point and the 100% end of the white power scale, indicating full RF power output. The ALC should be showing one or two blue bars at the most. Any more than that and the ALC will be controlling the transmitter's output level which is not desired for digital modes. Keep reducing the <USB MOD Level> until the ALC reading is between zero and two blue bars. The low ALC setting ensures that nothing is being overdriven along the transmitter audio chain. The nearly 100% RF Power reading indicates that the digital mode is sending at full power. It should not be sitting continuously at 100%. Leave a little headroom.

- If you want to reduce power later, you can turn down the RF Power using the MULTI control or reduce the send level in the digital mode program.

- I ended up with the USB MOD Level set to 17%. Your setting may be different depending on the audio settings in your PC and digital mode software. I have adjusted all of my digital mode programs, MRP40, MixW, and WSJT-X so that they all transmit at the same level. i.e. full RF power and one or two bars of blue on the ALC meter.

THE FT8 PRESET

The February 2021 (V1.30) firmware update added a "one-touch FT8 Preset mode" to the second page of the main menu. It adds five one-touch "mode pre-sets." The idea was to quickly change the radio to the settings required for FT8, reflecting the popularity of the mode. The top pre-set is labelled 'Normal,' but I have renamed it to 'SSB' because it does not return the radio to the previous non-Preset setting. The second item is labelled 'FT8,' although the options that it sets are suitable for most external digital modes. You can edit those two settings and add three more pre-set arrangements of your own design.

Load the FT8 Preset before you start WSJT-X or another digital mode program. <MENU> <2> <PRESET> <FT8> <YES>

Unload it again when you have finished with the FT8 or other Preset. <MENU> <2> <PRESET> <UNLOAD> <YES>

Tip: one of the things that you can add to a 'Preset' is the Icom CI-V address. This is very handy if you have digital mode software that does not support the IC-9700 and needs a different CI-V address. Simply make a 'Preset' profile with the CI-V address that the digital mode software expects. For example, the IC-7300 uses 94h. When you 'Unload' (turn off) the 'Preset' the CI-V address will return to the normal A2h setting.

When you have finished using the Preset, touch the 'UNLOAD' Icon. Do not select 'Normal' because that is just another Preset, and it may well be different from your normal settings.

> Activation rules

You cannot load a preset with the preset mode set to DD unless the radio is on the 1200 MHz band, because the radio can only transmit DD (digital data) on that band.

You cannot activate any Preset while:

- Transmitting in the DV mode
- The radio is in DR mode
- The DV Gateway mode is active
- The radio is in Satellite mode
- You have selected a blank memory channel
- You also cannot load a preset when the preset mode is DD unless the radio is on the 1200 MHz band, because the radio can only transmit in the DD mode on that band.

> To add a Preset

You can save the current radio settings as a new 'Preset,' <MENU> <2> <PRESET>. Touch and hold a vacant 'Preset' slot, or you can overwrite an existing 'Preset. Then select <Save to the Preset Memory>.

Add a Preset Name. Make any other changes you need, then scroll down to page six and select <<Write>> <YES>.

> Editing the Preset settings

You cannot edit a 'Preset' if it is in use. Touch UNLOAD first, <MENU> <2> <PRESET> <UNLOAD> <YES>.

Touch and hold the 'Preset' slot you wish to edit. Then select <Edit the Preset Memory>. Make any other changes you need, then scroll down to page six and select <<Write>> <YES>.

Each PRESET can store, a Preset name, the mode, receiver filter, filter bandwidth, USB keying settings, CI-V settings, data mod type, data off mod type, TX bandwidth, speech compressor, and TX wide/mid/narrow.

You can select items that you want to store and unselect any irrelevant items. For example, I set up a Preset for FM receiving. It does not need any of the CI-V or transmitter settings.

➤   Turning off the Preset

A Preset set on one band will not apply if you change to a different band. But if you return to the band with the Preset, it will still be in place.

If you change any of the settings, such as changing mode, the radio automatically unloads the Preset. If you change the mode back again, the Preset settings will return. The permanent way to turn off a Preset is to use <MENU> <2> <PRESET> <UNLOAD>.

➤   Preset settings

Adding a preset automatically saves the following settings. They cover pretty much everything you are likely to change when switching to digital mode operation.

Name of the preset	USB IF Output Level	USB Keying (RTTY)
Mode	USB MOD Level	USB Inhibit Timer
Filter	DATA MOD	CI-V Baud Rate
Filter BW (144 MHz)	SSB-D TX Bandwidth	CI-V Address
Filter BW (430 MHz)	DATA OFF MOD	CI-V Transceive
Filter BW (1200 MHz)	COMP	CI-V USB Port
Filter type	SSB TBW	CI-V USB Baud Rate
USB Output Select	SSB TX Bandwidth	CI-V USB Echo
USB AF Output Level	USB SEND	
USB AF SQL	USB Keying (CW)	

SETTING UP FOR EXTERNAL DIGITAL MODE SOFTWARE

To use a PC software digital mode program for JT65, FT8 or any other external digital mode program, you will first have to establish communication between the PC and the radio. You can use the old CI-V interface via the REMOTE jack and send audio through an interface box to the radio, but that is terribly old-fashioned, not to mention relatively difficult. So, I will describe the easy way, using a USB cable.

You will also have to set up an audio connection and set the audio levels. These steps are covered in the previous sections. Generally, this only needs to be done once and after that, any differences between PC programs can be managed by changing settings on the PC.

*Note that audio is sent to the PC when it is in any mode, so you can use your digital mode PC software to see and decode the digital mode signals. But the radio **must** be in the USB-D data mode to accurately transmit digital mode signals from your PC.*

In the non-data voice modes, if you have the default DATA OFF MOD setting of MIC, ACC the external software will be able to operate the PTT, but no audio will be passed to the modulator. If you have changed DATA OFF MOD to MIC, USB then audio will be passed through, but the microphone will be live, the transmit audio filters will be incorrect, and VOX or Compression may be enabled.

➢ First steps

First set up a connection between the radio and the PC. Follow the steps in the 'Setting up a connection with your PC' section starting on page 23.

Then set the Windows PC sound settings on page 24 and the USB port audio settings on page 31.

That's it! You are finished with the setup.

➢ RF Power in digital modes

I have not experienced any problems running 100% power on digital modes. However, after setting up the radio for digital mode operation, I suggest monitoring the multi-function temperature meter for a few overs just to make sure that everything is OK. Touch and hold the meter scale or press <MENU> <METER> to show the multi-function meter. The temperature meter is bottom right. If the temperature meter does not get up to 'hot' you can run 100% transmit power. If the transmitter is getting hot, reduce the RF power to 75% or 50% by turning the Multi knob.

➢ TBW setting for FT8 mode

If you are planning to use the radio for the FT8 mode, the transmitter bandwidth (TBW) setting doesn't matter because the WSJT-X program will use the Split mode to offset the transmit signal. There is no need to widen the transmit bandwidth.

SETTING UP THE RADIO FOR RTTY OPERATION

The radio supports three kinds of RTTY operation. Firstly, there is the onboard RTTY decoder. Which can be used with the RTTY message memories.

In this mode, you can take advantage of the excellent TPF (twin passband filter). I recommend using the TPF filter all the time because it really helps with accurate decodes. The second method is to use external PC software such as; MixW, MMTTY, MMVARI, Fldigi, etc. with AFSK (audio frequency shift keying). AFSK uses two audio frequencies to create the frequency-shift keying in the SSB mode.

The third method is the FSK mode which uses a digital signal to key the transceiver to predefined mark and space offsets.

➢   Onboard RTTY operation

Onboard RTTY operation centers around the built-in RTTY decoder and the eight message memories. Unfortunately, this radio has the same problem as the IC-7300. You cannot connect an external ASCII keyboard to the radio, so you can't enter a station's callsign or any other text. For contests, you can edit the keying macros during every QSO but that is not very satisfactory. Even an onscreen keyboard could create this essential function. So, while the RTTY decoder works very well, the built-in RTTY function is not very usable.

➢   RTTY mode settings

Select the RTTY mode. If the radio is already on RTTY it will select RTTY-R. This changes the operating sideband and more importantly it reverses the Mark and Space frequencies. If decode is gibberish the other station might be transmitting on the other sideband, creating RTTY-R.

You can change the RTTY MARK tone frequency and the RTTY shift. The default is that the Mark tone is at 2125 Hz and the Mark/Space shift is 170 Hz. In RTTY-R the Space tone is at 2125 Hz and the Mark/Space shift is still 170 Hz.

You can change the settings, but I don't see any reason to do so.

- <MENU> <SET> <Function> <RTTY Mark Frequency> 1275, 1615, or 2125
- <MENU> <SET> <Function> <RTTY Shift Width> choose 170, 200, or 425
- <MENU> <SET> <Function> <RTTY Keying Polarity> Normal or Reverse

➢   Keyer send messages (keyer macros)

There are eight RTTY message menus which can be used for DX or Contest operation. They are great for sending CQ on a satellite. At a minimum, you will have to add your callsign, or the macros will send 'Icom' in place of it.

To edit the messages. Put the radio into RTTY mode and select <MENU> <DECODE> <TX MEM> <EDIT>. Touch the message that you want to edit. Then select EDIT.

An on-screen keyboard will appear, and you can edit or replace the message. Each message can be up to 70 characters long.

The ↵ symbol is a carriage return. It creates a new line on the text display at the receiving station.

Make sure that you press ENT, or your changes will be lost when you exit the keyboard screen.

Touch the Return icon ↺ or the EXIT button to exit each menu layer.

To send the messages, while in the RTTY mode, select <MENU> <DECODE>. Then <TX MEM> and a message RT1 to RT8. After the message has been selected, the display reverts to the decode screen.

➢ RF Power in RTTY mode

You can run 100% power on RTTY, but if you are prone to very long 'overs' I suggest that you keep an eye on the temperature meter on the multi-function meter display. If the transceiver is running very hot, de-rate the transmitter power. Press MULTI select the RF POWER icon and reduce RF power to 75% or 50% by turning the Multi knob.

➢ AFSK RTTY audio levels – external program

When using the built-in RTTY mode the modulation level is not user adjustable. If you are using an external digital mode program the levels should be set as per the Windows PC sound settings on page 24 and the USB port audio settings on page 31. You can set the audio levels in the IC-9700, or you can set the levels using the soundcard mixer controls in your PC or possibly with level controls in the digital mode software.

I have chosen to leave the audio level being sent to the PC at the default 50% setting and will adjust the audio levels on the PC if required. If you want to change the audio level being sent to the PC, Select <MENU> <SET> <Connectors> <USB AF/IF Output> <AF Output Level>.

I have chosen to reduce the audio level being used to modulate the transmitter so that the RF output power is just reaching 100% and the ALC is not more than two blue bars. That ensures that nothing is being overloaded along the audio chain.

To change the audio level from the PC to the transmitter, select <MENU> <SET> <Connectors> <MOD INPUT> <USB MOD Level>. Mine is set to 17%.

➢ FSK RTTY from an external PC program

For RTTY I would normally use the MixW PC software running RTTY in AFSK mode, rather than using FSK. I have not found a way to get MixW to operate the radio in FSK mode. But you can create FSK RTTY using MMTTY. One advantage of using FSK rather than AFSK from an external PC program is that you use the RTTY mode on the radio rather than the USB-D data mode. That means that you can use the TX MEM messages and the excellent TPF (twin peak filter).

First set up a connection between the radio and the PC. Follow the steps in the 'Setting up a connection with your PC' section starting on page 23. Then set the Windows PC sound settings on page 24 and the USB port audio settings on page 31.

If you want to run FSK RTTY and set up with the N1MM logger using MMTTY, refer to the notes by K0PIR at http://www.k0pir.us/icom-7300-rtty-fsk-mmtty/ or a video on the topic at https://www.youtube.com/watch?v=NmNHVjjAdiY.

➢ IC-9700 Setup for FSK RTTY

The FSK keying must be on a different COM port to the CI-V control. Set <MENU> <SET> <Connectors> <USB SEND/KEYING> <USB Keying (RTTY) to USB (B) DTR

➢ MMTTY Setup for FSK RTTY

1. On the MMTTY program click Option(O)

2. Select Setup MMTTY(O)

3. Click the TX tab and then the Radio command button.

4. Set the Port to the port that is normally used for CI-V control (Com9).

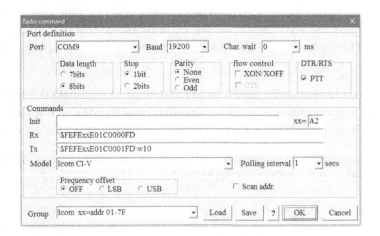

5. Back on the TX tab select EXTFSK64 in the PTT & FSK dropdown box. This is the Windows 10 setting. For 32 bit Windows or other 32 bit operating systems use the EXTFSK option.

6. Selecting the EXTFSK64 (or EXTFSK) option will open a small popup window. Sometimes it ends up under other windows especially if you neglected to close it previously. If you can't find it click on the MMTTY icon on the bottom toolbar and you should see it hiding, there. Set the Com port that is to be used for FSK operation. In

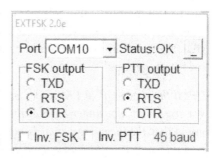

my case, it will be COM10. This cannot be the standard CI-V port. It must be the second Com port. Set output PTT to RTS and FSK output to DTR. This must match the CI-V settings in the radio. **Do not forget to close this little popup window by clicking the box next to Status OK: You may have to close other MMTTY windows first.**

7. On the Misc tab change TX Port to COM-TxD(FSK). This makes the application use the DTR Com port control line to send a digital FSK signal rather than 'Sound' which sends AFSK tones.

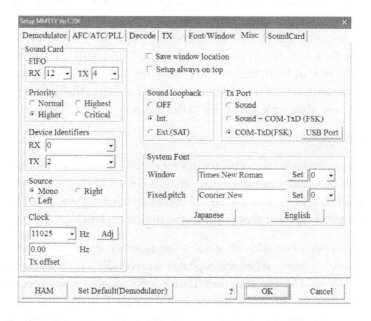

SETTING UP FOR FM OPERATION

There is no special setup for the microphone gain on the FM mode. It can be left as it was set for SSB. Or set anywhere up to 50%.

Repeater operation is usually achieved by selecting repeaters that you have saved into the 99 memory channels per band. These can be accessed by pressing the V/M button to change to MEMO mode and stepping through the channels with the MULTI knob. The scan function can be used to check the whole bank or one of the three scan groups.

TIP: If the receiver is in FM mode, is receiving a signal, and the squelch is open, there will be a green indicator under the S at the left edge of the S meter. If the indicator is flashing it means that the incoming signal is not on the frequency that the VFO is set to. Either the incoming signal is off frequency, or the receiver is not tuned to the correct frequency.

➢   Auto Repeater – US and International models

If you tune the VFO across the repeater output section of a band while in FM mode, the radio can automatically select the correct duplex split so that you will transmit on the repeater input frequency. This could be very helpful if you don't know the frequencies of the repeaters in the area. You simply tune until you hear the repeater output signal and transmit. It also saves from you having to manually set the duplex option if you save the channel to a memory slot. Apparently, this feature is only available on the US and International variants of the radio. Check <MENU> <Function> <Auto Repeater> <ON (DUP, TONE)>. As far as I can tell you can't set the frequencies at which the auto-duplex function operates, from within the menu structure. But you can edit the 'Auto Repeater Edge' data table in the CS-9700 programmer or the RT Systems programmer on the 'Band Settings' tab of the 'Radio Settings Menu' and upload the revised data to the radio. I recommend you download the current state of the radio, make any changes to the data table, save the file on your PC, and then upload it back to the radio.

➢   Setting the TONE

TONE squelch is often used for repeater operation and some FM satellites require a tone on the uplink signal. The radio supports CTCSS, an analog tone coded squelch system and DCS (DTCS) which is a digital coded squelch system. It is unlikely that you will encounter a repeater that requires the digital system.

Rather than the repeater squelch or 'gate' opening when the input signal exceeds a pre-set signal level, it opens when the repeater receiver detects the sub-audible tone being sent with your FM transmission. Using tone squelch stops repeaters being constantly triggered by noise bursts or interference. Usually, the same tone is retransmitted by the repeater, so you can use tone squelch on your receiver as well.

You can access the Tone settings by pressing the TONE RX-CS button, or via the <FUNCTION> <TONE> Soft Key. The TONE Soft Key is only available when the transceiver is in FM mode. A blue indicator around the Soft Key indicates that tone squelch is in use.

There are three CTCSS tone modes; OFF, TONE, and TSQL. You can ignore the DTCS modes since it is unlikely that you will need to use DTCS.

- The TONE mode sends a tone with your FM transmission that opens the repeater squelch.

- The TSQL mode squelches your receiver until the correct tone is received from the repeater. It also sends the tone on your transmission to open the repeater squelch.

- The OFF mode turns off both receiver tone squelch and the transmitted tone.

Touch and hold the TONE RX-CS button or the TONE Soft Key to open the setup screen.

- TONE is the tone that is sent to the repeater in the TONE mode.

- T-SQL is the tone that must be received from the repeater to open the receiver squelch in the TSQL mode. The same tone is sent with your transmit signal.

- DTCS is the code number that would be sent if DCS (DTCS) was being used.

The full set of CTCSS tones is available. Most repeaters use either 88.5 Hz or 67 Hz.

Touch TONE or T-SQL and use the main VFO knob to change the tone. Touch and hold DEF to reset the tone back to the default of 88.5 Hz.

CTCSS Tones (Hz)				
67.0	69.3	71.9	74.4	77.0
79.7	82.5	85.4	88.5	91.5
94.8	97.4	100	103.5	107.2
110.9	114.8	118.8	123.0	127.3
131.8	136.5	141.3	146.2	151.4
156.7	159.8	162.2	165.5	167.9
171.3	173.8	177.3	179.9	183.5
186.2	189.9	192.8	196.5	199.5

CTCSS stands for, 'Continuous tone coded squelch system.' A sub-audible tone is sent with your transmission to open the receiver squelch at a repeater. It may also be transmitted by the repeater to perform the same function in your receiver. CTCSS was originally developed to establish talk groups within commercial mobile radio fleets. By selecting a tone, you could select the mobiles or handheld radios that would hear your transmission. Change the tone to call a different group of mobiles.

There is a useful icon called T-SCAN on the TONE FREQUENCY sub-menu.

- To find the tone that you should be transmitting for the TONE mode. Tune your receiver to the repeater input frequency then touch the REPEATER TONE icon. When another station is using the repeater, touch T-SCAN. The radio will scan through all the possible CTCSS codes until it finds the one that matches the tone on the other station's repeater input transmission.

  This method will usually work if the receiver is tuned to the repeater output frequency as the same tone is usually transmitted from the repeater.

- To find the tone that you need for tone squelch in the TSQL mode. Tune your receiver to the repeater output frequency then touch the T-SQL - TONE icon. When another station is using the repeater, touch T-SCAN. The radio will scan through all the CTCSS codes until it finds the one that matches the tone on the repeater's output transmission.

➢ Memory channels and FM

There is no specific memory storage for FM channels. You can use the standard memory channels and name the memory channel with the repeater identification. You can store FM channels in the DV/DD memory slots as well and access them while in the DR mode. See the D-Star chapter for more details. I prefer to use the standard memory banks for FM channels and reserve the DV/DD memory slots for digital voice repeaters.

Recalling a stored FM channel is the same as for any other saved memory. Press V/M to change to the memory channel mode and step through the channels using the MULTI knob or the Up and Down buttons on the microphone. Or Press <MENU> <MEMORY> and use the UP DOWN Soft Keys or the MULTI knob to select a channel.

The easiest way to set up your repeater channels is to use the CS-9700 or the RT Systems memory manager and enter all of the local repeaters into the memory slots in an organized way. Then upload the data back to the radio.

➢ Tuning to and saving a repeater channel

To store a repeater channel into a memory slot. Set the VFO to the repeater output frequency. If you have the Auto Repeater function the correct duplex split will be selected as soon as you tune to the repeater output frequency. If not, then select the appropriate duplex offset by pressing <FUNCTION> <2> <DUP>.

Touching DUP again will cycle through the DUP+, DUP-, and OFF settings.

The amount of duplex shift should be correct, but touch and holding the DUP soft key will let you change it if required.

Next set the CTCSS tone as described in 'Setting the TONE' above.

At this stage you should be able to trigger and use the repeater, so the next thing is to save the channel so that you can access it easily in the future. See 'Saving a frequency to a memory slot' on page 94.

USING AN SD CARD

You must have an SD card in the radio if you want to record signals off the air, store the radio configuration, store memory channel contents, RTTY decode logs, voice message memories, or saved screen capture images, or upload new firmware. You do not need an SD card for the CW or RTTY message keyers.

The blue SD indicator in the top right of the display just to the left of the clock indicates that there is an SD card in the SD card slot. A flashing SD icon indicates that the radio is writing information to, or reading information from, the SD card.

You can use a 2 Gb SD card or an SDHC card from 4 Gb up to 32 Gb. Icom recommends using SanDisk© SD cards. I am using a 16 Gb SanDisk micro SDHC card in an adapter, but an 8 Gb card is more than ample. I have data for three different Icom transceivers saved on to my SD card and the information screen says that I have 137 hours of recording time left. Unless I am transferring images to my PC, I leave the SD card in the radio all the time.

Icom recommends formatting the SD card using the function in the radio, before using it for the first time. <MENU> <SET> <SD Card> <Format>. There is also a menu option to 'unmount' the SD card before you remove it. (The same as you would when removing a USB stick from your PC). <MENU> <SD Card> <Unmount>. I don't bother using it and haven't had any problems, but it's your risk.

FIRMWARE UPDATES

Updating the firmware is a bit scary because during the process you are presented with some bright yellow warning screens. However, I found that updating the firmware went very smoothly.

You need an SD card and a PC that can write to it. If your PC can't write to an SD card you will have to buy a USB SD card reader. Mine cost about $10.

➢   Checking the installed firmware revision

The currently installed firmware revision is displayed at the bottom right of the splash screen during the radio startup process. Or you can find it at <MENU> <SET> <Others> <Information> <Version>.

The display indicates the six firmware versions. The overall 'firmware version' is the Main CPU revision number.

- Main CPU: (version 1.30 or newer)
- Sub CPU: (version 1.00 or newer)
- Front CPU: (version 1.00 or newer)
- FPGA Program: (version 1.07 or newer)
- FPGA Data: (version 1.00 or newer)
- DV DSP: (version 1.06 or newer)

➢   Downloading firmware from the Icom website

Visit the Icom website at https://www.icomjapan.com/support/firmware_driver/. Enter IC-9700 into the search box and look for 'Firmware.' Currently, it is version 1.30. If the latest release is newer than the one currently installed on the radio, you should download the new firmware. A single file is used to update all of the firmware updatable devices on the radio.

Click on the IC-9700 link and then check the "I have read, fully understand and agree to the above" checkbox. Then click on Download. After the download has finished, find the downloaded file in your computer's download directory. Right-click the filename and select 'Unzip.' Either unzip the file directly to the IC-9700 directory on the SD card or unzip it to your hard drive and then copy the file across to the IC-9700 directory on the SD card.

The unzipped file will be called something like '9700E113.dat.' It **must** be copied into the IC-9700 directory on the SD card, or the radio will not be able to find it.

Never turn the radio off during a firmware update.

➢ Firmware update process

- Unmount the SD card from the PC and plug it back into the radio. Wait a few seconds until the drive is 'mounted.' Blue SD icon not flashing.

- Select <MENU> <SET> <SD Card> <Save Setting> and save the current radio configuration. You can select a file name or just go with the default 'time and date stamp' file name. Touch the down arrow to get to page 2/3.

- Touch <Firmware Update> then after scrolling down and reading both pages of the yellow screen, be brave and select <YES>

- A list with at least one firmware file will be displayed. Select the file with the highest revision number. If a file is not displayed, it means that the firmware '.dat' file is missing or not in the IC-9700 directory.

- After you have read the scary precautions on the yellow screen, be brave again, and touch and hold <YES> for one second. You have already saved the radio configuration file in step two... right?

- Follow the instructions on the transceiver screen. **Do not turn the radio off** or touch any controls.

- The IC-9700 will read and check the firmware file from the SD card and then write the Main CPU and DSP/FPGA firmware updates to the radio. Progress is displayed on the screen.

- When the update is completed, "Firmware updating has completed." is displayed and the IC-9700 will automatically restart.

- As the radio boots, you will see the new firmware revision number displayed at the bottom right of the splash screen during the radio startup process.

- After the radio starts it should be back to normal operation. You can check out the new firmware revision numbers at <MENU> <SET> <Others> <Information> <Version>.

- Your settings will probably be intact but if they are not, you can recover them using <MENU> <SET> <SD Card> <Load Setting> and selecting the file name you saved earlier. Temporary settings such as the MPad memories and the band stacking registers may be lost.

SETTING THE CLOCK

The radio has a clock display in the top right of the display next to the SD icon. Generally, you set the clock to display the local time, but you could set it to display UTC (universal coordinated time).

Touching the clock icon opens a popup display of Local and UTC day, date and time.

➢ Setting clock to local time.

1. Select <MENU> <SET> <Time Set>

2. Select <UTC Offset> and use the plus or minus keys to set the UTC Offset for your location. Remember to add daylight saving time if daylight saving is in use. Touch the Return icon ↺ to exit.

3. Select <Date/Time>

4. Select <Date> use the up-down keys to set the year, month, and day (in that order). Make sure that you touch the SET icon when you have finished, or your changes will not be saved.

5. Select <Time> and use the up-down keys to set the current local time in 24-hour format. Make sure that you touch the SET icon when you have finished, or your changes will not be saved.

6. Press the EXIT button or touch the Return icon several times ↺ to exit.

➢ Setting the clock to UTC.

1. Select <MENU> <SET> <Time Set>

2. Select <UTC Offset> use the up-down keys to set the UTC Offset to zero (0).

3. Select <Date/Time>

4. Select <Date> use the up-down keys to set the UTC year, month, and day (in that order). The date may be different from the date at your place. Make sure that you touch the SET icon when you have finished, or your changes will not be saved.

5. Select <Time> and use the up-down keys to set the current UTC time in 24-hour format.

6. Press the EXIT button or touch the Return icon several times ↺ to exit.

NTP TIME SYNCHRONIZATION

NTP time synchronization is available if the Ethernet LAN port is connected to your home network and an Internet connection is available.

Select <MENU> <SET> <Time Set> <Date/Time> <NTP Function> <ON> and the clock will be synchronised to an Internet time server. The default server is time.nist.gov, which is in the USA, but you can change it if required.

It seems a bit pointless to dedicate an Ethernet connection just for keeping the clock display accurate, but you can. Alternatively, you can connect a GPS receiver. See GPS time synchronization.

GPS TIME SYNCHRONIZATION

If you have an external GPS receiver connected for D-Star, it can be used to keep the clock display accurate.

Select <MENU> <SET> <Time Set> <Date/Time> <GPS Time Correct> <Auto>

For information about connecting a GPS receiver, see 'GPS location' on page 180.

SETTING AMATEUR BAND LIMITS

The radio should not allow you to transmit outside the amateur radio bands, but these vary between IARU regions and from country to country. The radio beeps if you tune the VFO across an amateur radio band edge. To some extent, you can set these band edges to suit the legal requirements for your country. Basically, you can make bands smaller or split them up, but you can't add new ones. Note that these band edges are not the same as the ones that you set for the Panadapter 'FIX' mode.

If you live in the USA, your radio will already be set up for the American band plan and no changes will be necessary.

The 'User Band Edge' settings are hidden. To change the band limits from the factory defaults you first have to change the 'Band Edge Beep' setting. In the default ON setting, the 'Band Edge Beep' makes the radio beep when you tune to a frequency above or below an amateur radio band. You can turn off the beep using <MENU> <SET> <Function> <Band Edge Beep>.

To enable the User Band Edge menu option, you must set <MENU> <SET> <Function> <Band Edge Beep> to <ON (User)> or to <ON (User) & TX Limit>.

- The ON (User) option lets you change the frequencies where the Band Edge Beep occurs.

- The ON (User) & TX Limit option lets you change the frequencies where the Band Edge Beep occurs and block transmitting out of the band. Note that you must leave this option activated if you want the customized band edge to apply. If you revert back to any of the other settings, the hardware diode-based band limits will apply.

Once you have changed Band Edge Beep setting, you can select <MENU> <SET> <Function> <User Band Edge>. You will see a list of the three currently programmed amateur bands and the frequency limits. You can program up to thirty band edges, but you will probably only need three.

➢ To edit a band edge

The most likely reason to edit a band edge is to make the band conform to your local band plan. For example, the New Zealand 70 cm band extends from 430-440 MHz, but my radio was set for the US allocation and allowed transmission up to 450 MHz.

1. Touch one of the band edges and edit the lower and/or upper frequencies.

2. The software will not let you enter a frequency that is outside of the limits imposed by the hardware settings. In other words, you can make a band narrower but not wider.

3. Touch ENT to save your changes. Then Exit the screen.

4. The User Band Edge menu option must be left at 'ON (User) & TX Limit.'

5. Return to VFO mode and tune over the band limits. You should hear a beep as you tune over the band edge and you should not be able to transmit outside of the band.

➢ To split, add, or delete a band

There are some rules.

6. Touch and hold a band edge to Insert, Delete, or Set it to the Default setting.

7. You cannot have any band overlap any other band. If you want to split a band you will have to edit or delete the old band first.

8. You cannot enter a frequency that is outside of the pre-set amateur bands which are set in hardware with diodes.

9. The bands must be in order. Lowest frequency band first. So, if you insert a new band you have to do it in the right place. Insert adds a new slot above the one you touch and held. But you can only enter frequencies that are available. In other words, you have to have trimmed down an existing band to make room for your new one.

10. Are you still sure you want to do this?

Front panel controls and connectors

This chapter describes all the front panel controls. Generally, it provides some additional detail to supplement the information provided in the Icom manuals. Any menu options relating to the front panel controls are included.

POWER

I guess you have already worked this one out. During the boot-up process, the bottom right of the display shows the firmware revision. And if you have entered your callsign it will display under Icom IC-9700 on the splash screen. Next, you get a display stating, 'RF Power (band)' and the current RF Power setting for the band.

Pressing the POWER button for a short time can store a screen capture of the current display on to the SD card. Set <MENU> <SET> <Function> <Screen Capture [POWER] Switch> to <ON>.

You can select to save the image in .png or .bmp graphics format. I chose to keep the default .png setting. <MENU> <SET> <Function> <Screen Capture File Type> <PNG>.

Press and hold the POWER button for two seconds to turn the radio off.

TRANSMIT

The TRANSMIT button places the transceiver into transmit mode. Press it again to stop transmitting. Be careful to avoid transmitting full power for long periods of time. On the three voice modes, the microphone will be live while transmitting. On FM the transceiver will transmit 100 Watts. On AM it will transmit 25 Watts. On SSB it will transmit normally for SSB. The output power will be dependent on the microphone audio level. The TX indication on the main screen turns to white on red to indicate that the radio is transmitting.

In the DD (digital data mode) on the 23 cm band, the TRANSMIT button turns the transmit inhibit function on or off.

CALL/DR

Pressing CALL/DR toggles the radio between 'Call Channel mode' and the normal VFO or memory channel modes. You can store two call channels per band. Use the MULTI knob to switch between them. The CALL/DR button is a quick way to set the radio to the frequencies used by your favorite local repeater.

Holding down the CALL/DR button turns the DR (D-Star Repeater) function on or off. See D-Star DR Repeater mode on page 162.

VOX / BREAK-IN (VOICE MODES)

Pressing the VOX/BK-IN button while in a voice mode SSB, AM, FM, RTTY or one of the data sub-modes, will enable VOX (voice-operated switch) transmit switching. VOX keys the transmitter when you talk into the microphone without the need for you to press the PTT switch on the microphone. It is handy if you are using a headset or a desk microphone without a PTT switch.

Press and hold the VOX/BK-IN button to enable the VOX MULTI menu. There are four settings. Touch the required setting and turn the MULTI knob to change the value. You should take the time to set the VOX up carefully as some settings tend to counteract other settings.

Press the press EXIT button or touch the Return icon ↺ to exit.

➢ VOX GAIN

VOX GAIN sets the sensitivity of the VOX. In other words how loud you have to talk to operate it and put the radio into transmit mode. (Default 50%).

➢ ANTI-VOX

ANTI-VOX stops the VOX triggering on miscellaneous noise like audio from the speaker or background noise. Higher values make the VOX less likely to trigger. (Default 50%).

➢ DELAY

DELAY sets the pause time before the radio reverts to receive mode. It needs to be set so that the radio keeps transmitting while you are talking normally but returns to receive in a reasonable time after you have finished talking. (Default 0.2 seconds).

➢ VOICE DELAY

VOICE DELAY, (Default OFF), sets the delay after you start talking before the transmitter starts. Generally, you want the radio to transmit immediately to avoid the first syllable or word being missed from the transmission. However, if you are prone to making noises perhaps you should set it longer. If you start each transmission with "Ah" you could set it quite long.

VOX / BREAK-IN (CW MODE)

Pressing the VOX/BK-IN button while in CW mode cycles through semi break-in (BKIN), full break-in (F-BKIN), or no break-in modes.

See 'break-in setting' on page 20.

Press and hold the VOX/BK-IN button while in CW mode to enable the MULTI Break-in display. There is only one adjustment. Turn the Multi knob to change the semi break-in delay setting. The default is a period of 7.5 dits at the selected CW keying speed.

PHONES

The headphone jack is a 1/8″ (3.5mm) stereo mini-phone jack which provides audio output to both sides of a stereo headset or headphones. The impedance is a standard 8 - 16Ω. The main receiver output is on the connector tip and the sub receiver output is on the ring. The sleeve is for common ground. You could connect this jack to a set of amplified PC speakers. You can select an option that mixes the audio from the two receivers together. The best option is 'Auto' (default) which separates main and sub audio when both the receivers are active and sends the main audio to both sides when only the main receiver is active.

<MENU> <SET> <Connectors> <Phones> <LR/MIX ACC MOD LEVEL> <Auto>.

The audio output to the headphone jack is affected by the AF Gain (volume) control and the squelch controls.

MIC JACK

The Mic jack is for the microphone. There is no facility for the connection of a balanced microphone. The jack is a standard Icom 8 pin connector.

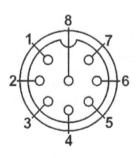

Pin	Description
1.	Microphone input (600Ω)
2.	+8 V DC output (max 10 mA) for electret microphones
3.	Up / Down buttons. 'Up' if grounded, 'Down' if grounded via a 470 Ω resistor. This pin can also be used to send message memories M1 to M4, see below.
4.	Squelch (goes low when squelch is open)
5.	PTT (pull low to activate transmitter)
6.	PTT ground
7.	Microphone ground
8.	Audio output (for 'speaker mics' the level is controlled by the AF volume controls). Menu select main or sub receiver.

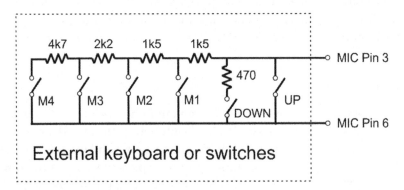

You can use the microphone jack to connect switches or a keypad used to send the first four of the eight keyer macros for the selected mode, (voice, CW, or RTTY). This might be useful if you have additional buttons on your microphone, or you wish to re-purpose the 'up' and 'down' buttons on the Icom Hand Mic.

To enable the external keypad or switches to trigger message transmission in the Voice, CW, or RTTY mode, you must set them to ON in the menu structure. This reduces the possibility of keying the transceiver by accident.

<MENU> <SET> <Connectors> <External Keypad> <VOICE>

<MENU> <SET> <Connectors> <External Keypad > <KEYER>

<MENU> <SET> <Connectors> <External Keypad > <RTTY>

CUSTOMIZED MICROPHONE BUTTONS

The UP and DOWN microphone buttons normally move the VFO frequency up or down by 1 kHz. If the Tuning Step function is on, the VFO moves by whatever the step size is, and in memory mode the buttons step through the memory channels.

Firmware release 1.30 (The February 2021) introduced the ability to change the function of the microphone buttons. You can allocate any of the 24 options to either of the two buttons. <SET> <Function> <MIC Key Customize>. The programmed buttons perform the same as the front panel buttons.

Function	Description
UP	(Default) 50 Hz step, or TS step, or memory channel
DOWN	(Default) 50 Hz step, or TS step, or memory channel
UP (VFO kHz)	1 kHz step, or TS step, or memory channel
DOWN (VFO kHz)	1 kHz step, or TS step, or memory channel
XFC	Hold to monitor transmit frequency (same as XFC button)

Function	Description
CALL	Selects the 'Call Channel'
VFO/Memo	Push to swap VFO to memory channel mode. Hold to copy a memory channel into the VFO
DR	Select the DR (D-Star) mode
FROM/TO	In the DR mode press to select the FROM or TO box
HOME Channel	Selects the HOME channel for the mode (VFO memory) or (DR memory)
BAND UP	Push to move up a band. Hold for the band stack
BAND DOWN	Push to move up a band. Hold for the band stack
SCAN	Press to start a scan. Press and hold for the Scan Menu
Temporary Skip	While scanning, press to skip a channel
SPEECH	Announce the frequency and mode, (S-meter optional). In the DR mode announces the DR callsign. In D-Star simplex mode, announces the frequency.
MAIN/DUAL	Press to swap main to sub. Press and hold to turn the sub receiver off
MODE	Push to select a mode. Press and hold to toggle USB to LSB, CW to CW-R or RTTY to RTTY-R
Voice/keyer/RTTY 1-4	Press to Send voice/CW/RTTY message 1-4 (dependent on the mode)
T-CALL	Transmits a 1750 Hz tone (EU version only)
RX>CS	In DV or DR shows history list. Hold to set the last calling station to the TO box
TS	Turn the tuning step mode on or off. Hold for TS screen
MPAD	Press to recall MPAD channel. Hold to save the current VFO settings to the MPAD
SPLIT	Press to toggle Split on or off. Hold for Quick Split.
A/B	Press to swap VFOs. Hold to save the hidden VFO settings to the currently displayed VFO

MAIN RECEIVER AF RF/SQL

The main receiver AF RF/SQL control is the upper pair of knobs on the left side of the radio. The black inner knob controls the AF Gain (volume) and the silver outer knob controls the receiver squelch and optionally the RF gain.

Press and hold the upper volume control knob to swap the upper (main) and lower (sub) bands over so that you can transmit on the other band.

Pressing the upper volume control knob moves the focus from the lower band to the upper band, the same as touching the upper half of the touch screen.

The silver outer knob can be configured in three ways. In the normal mode turning the knob clockwise from the '12 o'clock position' sets the squelch. Adjust it so that the receiver audio is muted when the receiver is only receiving noise. The green RX LED will go out. When a signal is received the squelch will open and you will hear the transmission. Turning the knob anti-clockwise from the '12 o'clock position' reduces the RF Gain of the receiver.

The squelch level is marked by a small white triangle at the top of the S meter. A green indicator under the S on the S meter indicates that the receiver squelch is open. If the indicator is flashing while you are in FM mode, it indicates that the squelch is open, but the incoming signal is off frequency.

There is also a LED to the right of the display next to the VFO knob. It will be green if either receiver squelch is open or red while you are transmitting.

To set the RF/Squelch mode select, <MENU> <SET> <Function> <RF/SQL Control>. The options are; AUTO, SQL, or RF+SQL. The default setting is RF+SQL. It works as described above. The AUTO mode acts as a squelch control for FM and AM operation and as an RF Gain control for SSB and the other modes. In SQL mode the control works as a squelch control only and RF Gain is set to maximum.

SUB RECEIVER AF RF/SQL

The sub receiver AF RF/SQL control is the lower pair of knobs on the left side of the radio. The black inner knob controls the AF Gain (volume) and the silver outer knob controls the receiver squelch and optionally the RF gain.

Press and hold the lower volume control knob to turn the second receiver on or off.

Pressing the lower volume control knob moves the focus from the upper band to the lower band, the same as touching the lower half of the touch screen.

The silver outer knob can be configured in three ways. In the normal mode turning the knob clockwise from the '12 o'clock position' sets the squelch. Adjust it so that the receiver audio is muted when the receiver is only receiving noise. The green RX LED will go out. When a signal is received the squelch will open and you will hear the transmission. Turning the knob anti-clockwise from the '12 o'clock position' reduces the RF Gain of the receiver.

The squelch level is marked by a small white triangle at the top of the S meter. A green indicator under the S on the S meter indicates that the receiver squelch is open.

If the indicator is flashing while you are in FM mode, it indicates that the squelch is open, but the incoming signal is off frequency.

There is also a LED to the right of the display next to the VFO knob. It will be green if either receiver squelch is open or red while you are transmitting.

To set the RF/Squelch mode select, <MENU> <SET> <Function> <RF/SQL Control>. The options are; AUTO, SQL, or RF+SQL. The default setting is RF+SQL. It works as described above. The AUTO mode acts as a squelch control for FM and AM operation and as an RF Gain control for SSB and the other modes. In SQL mode the control works as a squelch control only and RF Gain is set to maximum.

P_AMP / ATT (PREAMPLIFIER OR ATTENUATOR)

The P.AMP/ATT button controls the status of the internal preamplifier and the front-end attenuator. The button has the same function as the P.AMP/ATT Soft Key on the FUNCTION menu.

Pressing the button turns the Preamp on or off. The preamplifier gain seems to be about 12 dB on the 2 m and 70 cm band and 5.5 dB on the 23 cm band. Press and hold the P.AMP/ATT button to enable a 9 dB attenuator.

NOTCH (NOTCH FILTER)

Pressing the NOTCH button cycles between the automatic notch filter, the manual notch filter, and off. The manual notch filter is not available in FM mode. An indicator is briefly displayed in the center area of the screen. AN indicates that the Automatic notch is turned on, MN indicates that the manual notch filter is turned on.

Auto notch (AN) eliminates the effect of long-term interference signals such as carrier signals that are close to the wanted receiving frequency.

The manual notch (MN) can be placed anywhere across the receiver's audio passband, to reduce eliminate interfering signals. Holding down the Notch button opens the MULTI menu where you can select from wide, medium or narrow manual notch filters and set the notch frequency. Keep the filter as narrow as possible while effectively removing the interference.

➢   The audio scope

You can see the effect of the notch filter by opening the Audio Scope. Select <MENU> <AUDIO>. Tune to a frequency where you can hear a carrier 'birdie.' You must have the receiver squelch open to see signals on the audio spectrum scope. Turn on the auto notch and you will see the carrier signal disappear from the audio spectrum display and you won't be able to hear it anymore. Enable the manual notch and you will be able to see a black zone on the audio spectrum indicating a deep null in the signal. Changing from narrow to mid or wide makes the null zone wider. Adjusting the manual notch position moves the nulled band across the audio spectrum.

➤ Menu settings

There are two menu settings that can affect Notch operation. For AM and SSB you can select from Auto Notch only, Manual Notch only, or the default choice of both Auto and Manual notch options. Since the auto notch is best for interfering carriers and the manual option allows you to place the notch where you want it on the audio passband, I can't imagine why you wouldn't want both options. But you can change it if you want to. <MENU> <SET> <Function> <[NOTCH] Switch (SSB)> or <[NOTCH] Switch (AM)>.

NB (NOISE BLANKER)

Pressing the NB button turns the noise blanker on. On the display with the large VFO characters, Noise Blanker operation is indicated with a white NB indication just below the Hz digit of the VFO frequency. Other screens get a brief popup indication. Holding the NB button down brings up a sub-menu where you can adjust the Level, Depth, and Width. The noise blanker is designed to reduce or eliminate regular pulse-type noise such as car ignition noise. You may need to experiment with the controls when tackling a particular noise problem.

The LEVEL control (default 50%) sets the audio level that the filter uses as a threshold. Most DSP noise blankers work by eliminating or modifying noise peaks that are above the average received signal level. They usually have no effect on noise pulses that are below the average speech level. Setting the NB level to an aggressive level may affect audio quality.

The DEPTH control (default 8) sets how much the noise pulse will be attenuated. Too high a setting could cause the speech to be attenuated when a noise spike is attacked by the blanker. This could cause a choppy sound to the audio.

The WIDTH control (default 50%) sets how long after the start of the pulse the output signal will remain attenuated.

Set it to the minimum setting that adequately removes the interference. Very sharp short duration spikes will need less time than longer noise spikes such as lightning crashes.

The noise blanker is disabled when the radio is in FM mode.

Noise blanking occurs very early in the receiver DSP process. It is performed on the wideband spectrum before any demodulation or other filtering takes place. Noise reduction is performed on the filtered signal i.e. within the receiver passband.

NR (NOISE REDUCTION)

The noise reduction system in the IC-9700 is very effective. Pressing the NR button turns the noise reduction on for the active receiver. Holding the NR button down brings up a sub-menu where you can adjust the noise reduction level (default 5). Adjust the level to a point where the noise reduction is effective without affecting the wanted signal quality.

Noise reduction filters are aimed at wideband noise especially on the low bands rather than impulse noise which is managed by the noise blanker. The NR filter works best when the received signals have a good signal to noise ratio. By introducing a very small delay, the DSP noise reduction filter is able to look ahead and modify the digital data streams to remove noise and interference before you hear it.

SD CARD SLOT

You can use a 'full size' SD card to save data settings, including; audio recorded off the receiver, voice keyer messages, screenshots of the display, logs, the RTTY decode history, and the radio's current settings. It is also be used for firmware updates. Note that the CW and RTTY keyer messages and the memory slots are stored in the radio, not on the SD card.

It can be a 2 Gb SD card or any SDHC card from 4 Gb up to 32 Gb. Icom recommends using SanDisk© SD cards. I am using a 16 Gb SanDisk SDHC card. The information screen says that after recording all of the keyer macros and saving the settings a few times that I have 257 hours of recording time left. So, buying a 16 Gb card as I did is probably overkill. I leave the SD card in the radio all the time.

Icom recommends formatting the SD card or USB stick, using the function in the radio, before using it for storage. <MENU> <SET> <SD Card> <Format>.

There is also a menu option to 'unmount' the SD card or USB stick before you remove it. (The same as you would when removing a USB stick from your PC). <MENU> <SET> <SD Card> <Unmount>.

The blue SD icon at the top right of the display indicates that an SD card is installed. It flashes when the card is being written to or read from, rather like the hard drive LED on a PC.

MAIN DISPLAY SCREEN

The touch screen functions are covered in the next section, starting on page 67.

MULTI

Normally turning the MULTI knob adjusts either the memory channel (M-CH) or steps the VFO in 1 kHz steps. The option is selected by the kHz M-CH button. It is indicated by either a 'kHz' or 'M-CH' icon at the top right of the touch screen display. It can also be set to control RIT (receive incremental tuning), activated by the RIT button, and indicated by 'RIT,' or twin passband tuning (PBT), activated by the PBT button, and indicated by 'PBT.'

The MULTI knob can be made to become a semi-permanent control for a wide range of parameters. This is indicated by a white-on-orange icon in the top right of the touch screen display.

Pressing the MULTI button turns on a selection of adjustments that are relative to the mode that you are using. Touch the Soft Key icon to select the item that you want to change. Turning the knob changes the value of the selected setting.

Touching a Soft Key on the MULTI menu often turns a function on or off. Usually, when a function is turned on, a blue indicator appears to the left of the control. Soft Keys that have multiple options don't have the blue indicator. For example, the Notch Filter MULTI menu Soft Key cycles through the three manual notch width options; NAR, MID, and WIDE.

Pressing the MULTI button again turns off the selection screen. So, does touching the main display, pressing the EXIT button or the Return icon ↺.

➢ MULTI knob sub-menu items

Items marked with a star * can be allocated to the MULTI knob by touch and holding the adjustment icon.

SSB	SSB-D	CW	RTTY
RF Power *	RF Power *	RF Power *	RF Power *
MIC Gain *	MIC Gain *	KEY Speed *	TPF
COMP *		CW Pitch *	
MONITOR *	MONITOR *		MONITOR *
FM AM DV	DD	NB	NR
RF Power *	RF Power *	LEVEL *	LEVEL *
MIC Gain *	TX Inhibit	DEPTH *	

		WIDTH *	
MONITOR *			
NOTCH	VOX	BK-IN	TX PWR LIMIT
POSITION *	GAIN *	DELAY *	RF Power *
WIDTH	ANTI-VOX *		LIMIT
	DELAY *		
	VOICE DELAY		

➢ MULTI knob functions

You can allocate the MULTI knob to the following functions. Touch and hold controls refer to icons on the MULTI sub-menu.

Function	Control	Color / Notes
RIT	RIT button	Normal – all modes
kHz display	kHz M-CH button	Normal – all modes
M-CH	kHz M-CH button	Normal – all modes
PBT 1	PBT button	Normal – all modes
PBT 2	PBT button	Normal – all modes
RF Power	Touch and hold RF Power icon	Orange – all modes
MIC Gain	Touch and hold MIC GAIN icon	Orange – all voice modes
COMP	Touch and hold COMP icon	Orange – SSB only
MONITOR	Touch and hold MONITOR icon	Orange – all voice modes
KEY Speed	Touch and hold KEY SPEED icon	Orange – CW only
CW Pitch	Touch and hold CW PITCH icon	Orange – CW only
RTTY TPF	Touch and hold TPF icon	Orange – RTTY only
NB LEVEL	Touch and hold LEVEL icon	Orange – hold NB button
NB DEPTH	Touch and hold DEPTH icon	Orange – hold NB button
NB WIDTH	Touch and hold WIDTH icon	Orange – hold NB button
NR LEVEL	Touch and hold LEVEL icon	Orange – hold NR button
NOTCH	Touch and hold POSITION icon	Orange – hold NOTCH button
VOX GAIN	Touch and hold GAIN icon	Orange – hold VOX/BK-IN button
VOX ANTI	Touch and hold ANTI VOX icon	Orange – hold VOX/BK-IN button

VOX DELAY	Touch and hold DELAY icon	Orange – hold VOX/BK-IN button
BK-IN DELAY	Touch and hold DELAY icon	Orange – hold VOX/BK-IN button
TX PWR LIMIT	Touch and hold RF Power icon	Orange – TX PWR LIMIT

MENU, FUNCTION, M.SCOPE, QUICK, AND EXIT

The functions of the MENU, FUNCTION, M.SCOPE, QUICK, and EXIT buttons are complicated. They each have their own chapters following on from the chapter about the Touch Screen.

XFC

XFC stands for 'transmit frequency check.' It is quite a useful feature. It lets you check and listen to the frequency that you will transmit on. In the normal non-split mode pressing XFC opens the receiver squelch so that you can hear the frequency that the VFO is set to. In the split mode, pressing XFC opens the receiver squelch on the transmit (split) frequency. It also displays the split offset in kHz. Hold down XFC and turn the VFO knob to change the split offset.

If you are using RIT (receive transmit incremental tuning) pressing XFC will show you the actual frequency that you will transmit on. In the CENT screen mode, the whole panadapter will shift so that the transmit frequency is in the center and the orange T marker will be displayed. In the FIX screen mode, the orange T marker will be displayed at the transmit frequency.

TX/RX LED

A green light indicates that the radio is receiving, and the squelch is open on at least one of the receivers. The green indicator on the S meter shows which receiver it is.

No light indicates that the radio is receiving and the audio is squelched on both receivers.

A red light indicates that the radio is transmitting.

AUTO TUNE

You might think that the AUTO TUNE button has something to do with the antenna tuner, but it does not. It operates in CW mode to pull the receiver onto the correct frequency while receiving a CW signal.

It adjusts the VFO frequency until the received CW signal is at the tone set by the CW PITCH control. At that point, the received CW signal is exactly netted with the transmit frequency.

SPEECH / LOCK

Press SPEECH 🔑 to activate the Speech function. A female voice tells you the current S meter reading [optional], frequency digits, and mode.

There are a range of options under <MENU> <SET> <Function> <SPEECH>. You can choose from English or Japanese language. Other languages may possibly be available in other countries. You can set the speed of the delivery. You can set whether the message includes the S meter reading or not, and you can set the volume of the announcement. The MODE SPEECH setting selects whether you want an automatic announcement every time you change modes.

Press and hold SPEECH 🔑 to lock the VFO. This locks the main tuning knob and stops the radio jumping to another frequency if you accidentally bump the main tuning knob. This function may be useful if you are operating as a 'Run' station in a contest or you are a DXpedition station or a Net controller.

If the <Lock Function> is set to the MAIN DIAL, the rest of the touch screen is unaffected. You can still change the focus of the VFO from A to B and operate the V/M button although the frequency remains locked and you can activate split. During split operation, the Split Lock function can be used. Select <MENU> <SET> <Function> <SPLIT> <SPLIT LOCK> ON. With Split Lock set to ON, you can adjust the split transmit frequency in the usual way by holding down the XFC button and tuning the main VFO knob. With Split Lock turned off you can't adjust the transmit frequency in this way while the VFO is locked.

If <Lock Function> is set to PANEL, the whole display screen is locked as well as the main VFO. The panadapter and spectrum display stay working but the VFO switch V/M and split are disabled.

You can reverse the Speech and Lock functions so that Speech becomes the 'press and hold' function, and Lock becomes the 'press' function. Select <MENU> <SET> <Function> [SPEECH/LOCK] Switch and change from SPEECH/LOCK to LOCK/SPEECH.

RIT

RIT is 'receive incremental tuning.' It shifts the Main receiver frequency by up to ± 9.99 kHz without changing the VFO frequency or therefore your transmit frequency. It is usually used to tune in a station that is transmitting a little off frequency, but it can be used as a method of operating split. The amount of RIT offset is shown on the right of the screen at the top of the panadapter display. Use the MULTI knob to adjust the RIT offset. Press and hold RIT to clear the RIT Offset. Press RIT again to turn off the RIT function.

If the panadapter is in FIX mode, the green R marker shows the actual receive frequency including the RIT offset. Pressing XFC shows the T marker indicating the transmit frequency and it also lets you listen to the frequency that you will transmit on.

If the panadapter is in CENT mode, the line at the center of the panadapter shows the actual receive frequency including the RIT offset. The whole panadapter jumps to the transmit frequency when you press XFC or transmit.

KHZ/M-CH

In the VFO mode, pressing the kHz/M-CH button allows you to use the MUTI knob to adjust the current VFO frequency in 1 kHz steps.

Press and hold the kHz/M-CH button allows you to use the MUTI knob to step through the memory channel slots.

PBT

Twin passband tuning is a standard feature of Icom receivers. It allows the operator to shift or narrow the IF passband to reduce the effect of very close interference signals. Pressing the PBT button selects PBT1. Indicated in the top right of the screen. Press it again to select PBT2. While in the SSB, CW, or RTTY modes PBT is adjustable in 25 Hz steps. In AM mode PBT is adjustable in 100 Hz steps.

Use the MULTI knob to adjust the passband shift value. A small popup display indicates the changes you are making. Press and hold the MULTI knob to clear the PBT offsets.

The received RF signal passes through two filters which normally have identical passband responses.

Adjusting the PBT1 setting using the MULTI knob shifts the passband of the PBT1 filter higher or lower in frequency. Adjusting the PBT2 setting using the MULTI knob shifts the passband of the PBT2 filter higher or lower in frequency. The effect that this is having on the IF passband is indicated in a pop-up diagram on the display. The popup disappears very quickly after stop adjusting the PBT controls. But there is a way to make visible for longer. Touch and hold the filter icon on the display to open the filter setting window. This gives a better display of the filter settings and it doesn't disappear all the time.

After adjustment, a tiny white dot to the right of the filter icon indicates that PBT is in use.

Press and hold the MULTI knob to clear the PBT offsets. Press the press EXIT button to exit this screen.

SPLIT

Pressing SPLIT engages split mode. Split operation in the IC-9700 is different from many transceivers where a pre-defined 5 kHz or similar offset is applied. In the IC-9700 using split means that the transceiver will receive on the frequency indicated by the 'A' VFO and transmit on the frequency indicated by the 'B' VFO, (or vice versa). That might be on a completely different part of the band and using a different mode, so you do have to be careful.

Holding SPLIT down can have two different effects depending on the split menu setting. If QUICK SPLIT has been set to ON, holding down the SPLIT button sets the split nominated in the Split Offset menu setting. If QUICK SPLIT has been set to OFF, holding down the SPLIT button has no effect.

To change the split settings, select; <MENU> <SET> <Function> <SPLIT>.

For more information on how to use the split mode effectively, see 'Operating split' in the chapter on 'Operating the radio,' (page 84).

A/B (VFO A TO VFO B)

The IC-9700 has two VFOs for each of the two receivers. They are named 'VFO A' and 'VFO B.' The A/B switch selects which one is currently active. They are not active if the radio is in MEMO mode.

The twin VFOs are usually used for Split operation. The alternate VFO will be used as the transmit frequency when you press SPLIT. You need to be aware that the second VFO may be on another part of the band or set to a different mode. Check before you transmit.

V/M (VFO TO MEMORY MODE)

The V/M button changes the radio from VFO mode to the MEMO memory channel mode. To select a memory channel, you can either change to memory mode and then step through the channels using the MULTI knob or you can step to the channel that you want and then press V/M. In memory mode, the VFO A or VFO B indicator is replaced by MEMO. The touch and hold function sets the currently selected VFO to the frequency that is stored in the memory slot. But you don't have to do that to use the channel.

SCAN

Pressing the SCAN button displays a Scan sub-menu showing the three pre-set scan ranges for the current band and a delta scan option. Press and holding the SCAN button restarts a previously selected scan. Note that this is only a sub-set of the Scanning functions that are available by pressing MENU and selecting SCAN.

If you are in MEMO mode the Scan submenu changes to a choice of scanning all of the memory channels for the current band, or one of the three scan groups *1, *2, or *3. You can also scan all three scan groups, a delta scan between frequencies, or all channels in the current MODE. This will save the radio stopping on SSB slots when you want to scan FM channels.

The main MENU has different scan options and settings. See page 90.

The SCAN mode also works when you are in the DR D-Star digital repeater mode. It scans the DD/DV memory bank for repeaters within the range of the radio. See D-Star (DR) Scan Mode on page 176.

TONE/RX-CS

In FM mode the TONE/RX-CS button displays the Tone sub-menu. European version radios have a function where holding down the TONE/RX-CS button and the PTT button simultaneously transmits a 1750 Hz tone.

Touch and hold the TONE/RX-CS button to set the tone frequency or operate the T-Scan mode. See 'Setting the Tone' on page 40.

In the DV mode, pressing the TONE/RX-CS button displays the RX History table. Holding down the button captures the latest received callsign as a temporary call destination. It captures the repeater callsign or the station callsign if you are operating simplex.

MPAD

Pressing the MPAD button sets the VFO to the last stored memory pad frequency and mode. Pressing repeatedly cycles through the five (or ten) stored frequencies.

Press and hold the MPAD 'memory pad' button to save the currently active VFO frequency, mode, etc. into the memory pad. This is pretty handy if you want to remember a frequency and come back to it later, perhaps when the current QSO on the frequency has concluded.

The memory pad can store either 5 or 10 frequencies, depending on the setting, <MENU> <SET> <Function> <Memo Pad Quantity>. The memory stores the mode, frequency, filter setting, CW break-in, VOX, compressor, and AGC settings, but not the split offset.

Saving the active VFO frequency into a memory pad position moves the other frequencies down in the list. The previous contents of the bottom memory slot are discarded.

MAIN VFO KNOB

This is the main tuning knob. You know what it does. It is also used to set some sub-menu controls such as the filter bandwidth adjustment I like the tuning knob it has a nice weight. Under the dial, there is a lever to adjust the drag on the knob. I have never felt a need to adjust it.

CUSTOMIZED BUTTONS

You can change the functions of three of the front panel controls. The VOX/BK-IN button, the AUTOTUNE/AFC button, and the TONE/RX>CS button.

Use <SET> <Function> <Front Key Customize> to change the functions.

➢ VOX/BK-IN button

1. VOX/BK-IN - Default functions

2. CD - Press to display the call history

3. PRESET - Press to open the PRESET screen

4. Home CH - Selects the HOME channel for the (VFO memory) or (DR memory)

5. Temporary Skip - While scanning, press to skip a channel

6. Send voice/CW/RTTY message 1 (dependent on the mode)

7. Send voice/CW/RTTY message 2 (dependent on the mode)

8. Send voice/CW/RTTY message 3 (dependent on the mode)

9. Send voice/CW/RTTY message 4 (dependent on the mode)

➢ AUTOTUNE/AFC button

1. AUTOTUNE/AFC - Default functions

2. AUTOTUNE/AFC RX>CS – Default autotune function in CW, and turn on or off the AFC in DV, FM, or AM. In DV or DD mode press and hold to display the call history

3. TONE/RX>CS – acts the way the RX>CS and TONE button normally works. In FM mode it opens the Tone window.

Press and hold to open the Tone Frequency window. In DV or DD mode press to display the call history, press and hold to pop the last call into the TO box.

4. CD - Press to display the call history

5. CD/RX>CS - Press to display the call history. Press and hold to pop the last call into the TO box

6. PRESET - Press to open the PRESET screen

7. PRESET/RX>CS - Press to open the PRESET screen. Press and hold to display the call history

8. Home CH - Selects the HOME channel for the (VFO memory) or (DR memory)

9. Home CH/RX>CS - Selects the HOME channel for the (VFO memory) or (DR memory). Press and hold to display the call history

10. Temporary Skip - While scanning, press to skip a channel

11. Temporary Skip/RX>CS - While scanning, press to skip a channel. Press and hold to display the call history

12. Send voice/CW/RTTY message 1 (dependent on the mode)

13. Send voice/CW/RTTY message 2 (dependent on the mode)

14. Send voice/CW/RTTY message 3 (dependent on the mode)

15. Send voice/CW/RTTY message 4 (dependent on the mode)

➢ TONE/RX>CS button

1. TONE/RX>CS – Default setting. In FM mode it opens the Tone window. Press and hold to open the Tone Frequency window. In DV or DD mode press to display the call history, press and hold to pop the last call into the TO box.

2. AUTOTUNE/AFC – Autotune function in CW, and turns on or off the AFC in DV, FM, or AM.

3. CD/RX>CS - Press to display the call history. Press and hold to pop the last call into the TO box

4. PRESET/RX>CS - Press to open the PRESET screen. Press and hold to display the call history

5. Home CH/RX>CS - Selects the HOME channel for the (VFO memory) or (DR memory). Press and hold to display the call history

6. Temporary Skip/RX>CS - While scanning, press to skip a channel. Press and hold to display the call history

The touch screen display

The touch screen display is probably the most important feature of the radio. It provides controls that are not available as knobs or buttons on the front panel, access to the many menus and settings, the display of the operating parameters and of course the fabulous panadapter and waterfall display.

Having a high-resolution display allows the menu items to be presented in plain English rather than the cryptic codes and number systems used in older transceivers. This makes them much easier easy to understand. But there are a lot of settings available and they are spread over many screens. Which makes it hard to remember where to find particular settings. Different menu functions are accessed from the MENU, FUNCTION, QUICK, and MULTI buttons and the EXP/SET 'Soft Key' icon. Often the menu choices change depending on the operating mode you have selected.

According to the Icom manual, there are 33 different indicators on the LCD display. I won't bore you by listing each one. Instead, I will mention them throughout the book as part of my explanation of the radio's functions. The full list is in the Icom Basic manual on pages 1-4 and 1-5.

The following items are adjustable by touching the icons on the touch screen.

TIME

The time is displayed in the top right of the screen. Touching the time display opens a window displaying the local time and date and UTC time and date. The window stays open until you Press EXIT or touch the Return icon ↺ to exit.

METER DISPLAYS

Touch the meter scale to cycle through the transmitter meter options, or touch and hold the meter scale to display all the meters at once. The options are; Po (power out), SWR (standing wave ratio), ALC (automatic level control), COMP (audio compression), V$_D$ (power amp FET drain voltage), and I$_D$ (power amp FET drain current). On the expanded display with the larger panadapter and on the display with the small panadapter and no on-screen menu icons, the meter is rather small. But it works the same way.

➢ The receiver S-meter

While the radio is receiving, the meter always reads the received signal strength in S points. The squelch level is indicated by a small white triangle above the S meter scale.

➢ The multi-function meter

Touch and hold the meter scale or press the MENU button and touch the METER icon to display the multi-function meter. Touch and the multi-function meter again to close it. Or use the EXIT button.

The multi-function meter displays all of the transmit metering at the same time and also an indication of the temperature measured at the final transistors.

This display is ideal when you are setting up the transmit levels for SSB because you can monitor the power output, ALC reading, and compression at the same time.

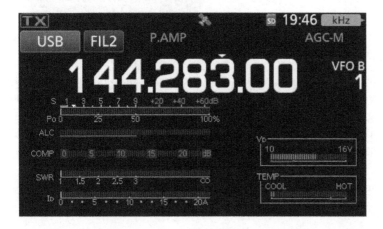

MODE

Touching the blue 'mode' icon brings up the MODE selection menu.

- SSB selects the correct SSB mode for the current band (USB or LSB) then exits the menu screen.

  o If the radio is already on USB, touching SSB will change the radio to LSB mode.

  o If the radio is already on LSB, touching SSB will change the radio to USB mode.

- CW selects the CW mode.

  o If the radio is already on CW it will select CW-R. This reverses the beat frequency oscillator so that it is higher than the received CW signal (CW-LSB mode), rather than the default (CW-USB mode).

  o If the radio is already on CW-R it will return the radio to CW.

- RTTY selects the built-in RTTY mode.

  o If the radio is already on RTTY it will select RTTY-R. This changes the operating sideband and more importantly it reverses the Mark and Space frequencies. If decode is gibberish the other station might be transmitting on the other sideband, creating RTTY-R.

  o If the radio is already on RTTY-R it will return the radio to RTTY.

- AM selects AM mode

- FM selects FM mode

- DV selects Digital Voice (D-Star) mode. This changes the DATA Soft Key into a GPS Soft Key. I don't know why they didn't just use the spare position.

- GPS allows you to select the data format (D-PRS or NMEA) for GPS data sent with your D-Star transmission, (if GPS TX is selected).

- DATA selects the Data mode where the transmit modulation is supplied over the USB cable from an external program, or over the ACC jack from an external device such as a Keyer, TNC, or PC.

  There are four data modes, USB-D, LSB-D, AM-D, and FM-D. The one that you are most likely to use is USB-D, the SSB data mode. You might possibly use FM-D for FM packet radio.

  To return to SSB from USB-D you have to touch DATA again rather than SSB. "Strange but true!" Touching SSB will not return you to USB. It will switch to LSB mode. You have to get into the mindset that the Data modes are subsets of the SSB, AM and FM modes. From SSB, touch data to get to USB-D and touch data again to exit back to USB. Why didn't Icom just add four more Soft keys for the four Data modes?

*Note that audio from the receiver is sent to the PC over the USB cable in all modes. But with the default Data Off Mod settings, audio from the PC will only modulate the transmitter if the radio is in a **data mode.***

CW can be operated from a PC while using the CW mode as it uses the RTS or DTR line for signaling rather than an audio tone. FSK (not AFSK) RTTY can be operated from a PC while using the RTTY mode as it uses the RTS or DTR line for data signaling rather than an audio tone.

FREQUENCY DISPLAY

The frequency display has several different touch screen functions.

Touching the inactive frequency display makes that VFO active so that you can change the functions of the selected receiver.

➢ MHz digits

Touch the MHz digit(s) to bring up the band change menu.

- Touch one of the three Soft Keys on the band you want to change to. The Soft Keys represent the band stacking menu.

- If a row of Soft Keys is greyed out, it means that the second receiver (or main receiver & transmitter) is using that band. You can' have both receivers active on the same band. Press and hold the lower volume control to turn off the second receiver and all three rows will become available again.

- Touch F-INP to enter a frequency directly into the VFO. You can also set a split offset or select a memory channel.

o ENT enters the data and closes the screen. It shortcuts the entry process. For example, type <1> <4> <5> <ENT> and the radio will set 145 MHz into the VFO.

o Touch a digit and then split to enter a split, e.g. <(-)> <5> <Split> sets a –5kHz split.

o If the radio is in memory mode, rather than VFO mode you can touch a number then MEMO to change to that memory channel. For example, <2> <1> <MEMO> changes to memory channel 21. Unfortunately, there is no indication of what is in any particular memory slot, so this function is a bit "hit and miss."

o Return ↺ closes the screen.

➤ kHz digits

Touch the three kHz digits to toggle the VFO tuning step between 'fast' and 'normal.' The 'normal' tuning step is either 10 Hz per step or 1 Hz per step, (see Hz digits below). A white triangle above the 1 kHz digit indicates that the VFO is in fast tuning mode. Touch and hold the three kHz digits to change the 'fast' tuning step (TS). You can set different 'fast' tuning steps for each mode. The default for SSB is 1 kHz but, I prefer 0.1 kHz (100 Hz). For FM you might select 12.5 kHz steps.

➤ Hz digits

Touch and hold the Hz digits to toggle between 'normal' 10 Hz per step tuning and 1 Hz per step tuning.

➤ Tuning rate

The tuning rate may speed up as you turn the main VFO knob. This is normal and is linked to the speed that you turn the knob. The feature is designed to get you to the other end of the band quickly when required. You can turn it to High, Low, or Off using <MENU> <SET> <Function> <MAIN DIAL Auto TS>.

➤ ¼ tuning speed

The digital modes and CW allow the use of the ¼ tuning function. Select <FUNCTION> <1/4> to turn the function on or off. The function slows down the tuning rate of the VFO to make tuning in narrow CW and digital mode signals easier. It is indicated with a ¼ icon to the right of the 10 Hz digit of the frequency display. If the 1 Hz tuning step is active it is indicated above the 1 Hz digit.

VFO mode display: In VFO mode, the VFO frequency is displayed in large white numbers. The currently selected memory position is displayed below the VFO A or VFO B indication on the right side of the display.

VFO mode

MEMO memory channel mode

Memory mode display: In memory mode, the line indicating the current VFO changes from VFO to MEMO. The memory slot number and scan group are displayed below. The name of the memory channel is displayed below the kHz and Hz digits of the frequency display. Press and hold the kHz M-CH button then turn the MULTI knob to change the memory slot. The small display with the panadapter visible does not include the name of the current memory channel.

SPECTRUM AND WATERFALL DISPLAY

Touching the screen within the spectrum scope/waterfall area expands a section of the display. Touching within the square moves the VFO to a frequency within the box. You will probably still have to use the main tuning knob to get the signal exactly tuned in. Touching the display anywhere outside of the box exits the expanded display with no effect on the VFO tuning.

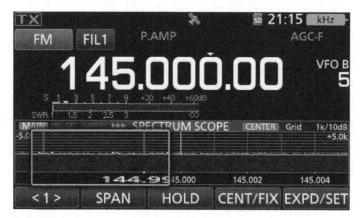

Selecting a frequency using the touch screen

The spectrum scope/waterfall display Soft Keys are only available with the medium or large panadapter displays. They are not available with the small panadapter display. This is a little annoying if you have the Voice message, Keyer, or Decoder screens open. You have to make sure that the panadapter is exactly how you want it before opening any of those other screens.

CENTER AND FIXED SPECTRUM DISPLAYS

The CENT/FIX Soft Key switches the spectrum display between the CENTER and FIXED spectrum displays. The 'fixed' display mode is better suited to observing a nominated band segment for activity. The 'Center' display mode is better suited to tuning across a band.

The February 2021 (V1.30) firmware update introduced a scroll mode to the FIX and CENT spectrum display modes. This is a good idea because it stops you tuning off the edge of the displayed spectrum. Touch and hold the CENT/FIX Soft Key to enter the SCROLL-C or SCROLL-F scrolling mode.

PANADAPTER AND WATERFALL DISPLAY

The panadapter can be used to display signals on the Main receiver or the Sub receiver if it is running. Touch the orange MAIN icon to change the panadapter to the Sub receiver. Touch the green SUB icon to change the panadapter to the Main receiver. *I don't think that this is mentioned in the manual. It's our secret!*

➢ Panadapter band edges (FIX mode)

When the panadapter is set to the FIX display, the panadapter acts like an SDR panadapter, displaying signals between pre-set frequencies.

The VFO frequency that the receiver is tuned to could be anywhere across the panadapter, or not even on the currently displayed panadapter. This mode allows you to monitor a predetermined section of the band. Touching the EDGE icon on the expanded display cycles through three panadapter bandwidth settings.

You can customize the panadapter band edges. For example, you might want one segment dedicated to the CW portion of a band, another to EME or meteor scatter, and a third to an FM repeater output segment. Each band allows three 'Fixed Edges' (panadapter bandwidths) with a maximum bandwidth of 1 MHz.

Note that the panadapter band edges are not the same as the amateur radio band edges that set the frequency ranges that you can transmit on and the band edge beeps that tell you when you have tuned outside of an amateur radio band. See page 47.

➢ Setting the panadapter band edges

If the Soft Key icons are not currently displayed at the bottom of the touch screen, hold <M.SCOPE> down for one second to enable the expanded screen.

Touch and hold <EXPD/SET> to open the 'Scope Set' menu and then select <Fixed Edges> at the bottom of the fourth setup screen.

Touch one of the three frequency ranges and then touch one of the three band edges. Set the lower frequency then press <ENT> and set the upper edge frequency. Remember to touch <ENT> to save your changes. The minimum panadapter bandwidth you can set for any segment is 5 kHz. The maximum is 1 MHz.

Touch and hold any of the three band edges to reset it to the Icom default.

➢ Scroll-F display mode

If you touch and hold the CENT/FIX Soft Key while the radio is in the FIX display mode, the FIX icon at the top of the spectrum display will change to SCROLL-F on a green background. Now, when you tune outside of the fixed frequency range the display will jump to the next band segment the same size as the fixed edge, and the receiver cursor will jump to the start of that segment, (tuning up or down). Selecting EDGE will change the display to a scan width equal to the width of the next fixed edge, but it will not change the VFO to the frequencies programmed into that fixed edge, i.e. the band edge segment sizes become the span size for the display.

➢ Panadapter span (CENT mode)

When the panadapter is set to the CENT display, the panadapter acts like a band-scope, displaying signals below and above the VFO frequency. The VFO frequency that the receiver is tuned to is always in the center of the panadapter. Touching SPAN changes the panadapter bandwidth.

➢ Scroll-C display mode

If you touch and hold the CENT/FIX Soft Key while the radio is in the CENT display mode, the CENT icon at the top of the spectrum display will change to SCROLL-C on a brown background. This switches the display to a FIX(ed) display with the span size set by the SPAN setting.

When you tune outside of the fixed frequency range the display will jump to the next band segment the same size as the Span setting and the receiver cursor will jump to the start of that segment, (tuning up or down). Selecting SPAN cycles through the span settings from ±2.5 kHz to ±500 kHz.

SPECTRUM SCOPE SOFT KEYS

The spectrum scope Soft Keys are displayed at the bottom of the touch screen. If they are not visible press and hold the M.SCOPE button.

The touch screen Soft Keys

➢ **< 1 >**

Change to Menu 2 selections.

➢ **SPAN**

When the panadapter is set to the CENT display, touching the SPAN icon on the expanded display cycles through eight panadapter bandwidth settings.

±2.5 kHz, ±5 kHz, ±10 kHz, ±25 kHz, ±50 kHz, ±100 kHz, ±250 kHz, and ±500 kHz.

Touch and holding SPAN sets the panadapter bandwidth to ±2.5 kHz.

➢ **EDGE**

In the FIX spectrum mode, EDGE cycles through three pre-set band edges.

➢ **CENT/FIX**

Toggles between the FIX and CENTER span modes. Touch and hold to enter the Scroll-F and Scroll-C modes.

- FIX mode displays the spectrum between pre-defined band edges. See EDGE.
- CENT displays the spectrum below and above the active VFO frequency. See SPAN.

- SCROLL-F mode displays the spectrum with a display span (width) defined by the pre-defined band edges. When you tune outside of the displayed bandwidth the display scrolls to the next band segment.

- SCROLL-C mode displays the spectrum with a display span (width) defined by the SCAN bandwidth. When you tune outside of the displayed bandwidth the display scrolls to the next band segment.

➢ HOLD

Freezes the spectrum scope and waterfall.

➢ EXPD/SET

Touch EXPD/SET to turn on or off the expanded spectrum scope display. This does not work if you have both receivers running and the lower VFO selected as active. Touch and hold EXPD/SET to access the SET menu.

➢ < 2 >

Change to Menu 1 selections.

➢ REF

Adjustment of the Spectrum display reference level. Annoyingly there is only one reference level control, so you often have to change the REF level if you change from the MAIN receiver to the SUB receiver and sometimes if you move from a narrow panadapter span to a wide one.

➢ DEF

Returns the Spectrum display reference level to 0 dB.

➢ SPEED

Cycle through FAST, MID, SLOW panadapter and waterfall speeds. The speed is indicated with blue markers on the top line of the Spectrum Scope.

➢ MARKER

Displays the orange (T) transmit frequency marker. In the FIX spectrum mode, the green (R) receive frequency marker is always displayed if the VFO is within the currently displayed panadapter bandwidth.

The full table of settings is in the SCOPE Spectrum Scope Soft Keys section on page 97. But there are a couple of scope settings that I believe most users will want to change immediately.

➤ CENTER Type Display

This is a very important setting. It sets the way that the wanted signal is displayed on the panadapter display. The Icom default setting is rather odd, placing the panadapter center in the middle of the filter passband instead of at the carrier point. I believe that most users will want to select either the 'Carrier Point Center (Abs Freq)' option or the 'Carrier Point Center' option. Touch and hold <EXPD/SET> <CENTER Type Display> and set it to <Carrier Point Center (Abs Freq)>. See page 100 for the full table of settings.

➤ Marker Position (FIX Type)

This is a very important setting. It sets the way that the wanted signal is displayed on the FIX mode panadapter display. The Icom default setting is rather odd, placing the panadapter center in the middle of the filter passband instead of at the carrier point. I believe that most users will want to select the 'Carrier Point' option. Touch and hold <EXPD/SET> <Marker Position (FIX Type)> and set it to <Carrier Point>. See page 100 for the full table of settings.

➤ Carrier point display options

- Filter center places the VFO frequency marker in the middle of the filter bandwidth, at the center of a USB or LSB signal. In my opinion, this is just plain wrong.

- Carrier Point (Center) places the VFO frequency marker at the carrier point. The left side of an upper sideband (USB) signal or the right side of a lower sideband (LSB) signal. In FIX mode bottom of the display shows the actual frequency. In CENT mode, the bottom of the display shows the span in kHz.

- Carrier Point Center (Abs Freq) places the VFO frequency marker at the carrier point. The left side of an upper sideband (USB) signal or the right side of a lower sideband (LSB) signal. The bottom of the display shows the actual frequency in MHz.

SPECTRUM SCOPE SET MENU

I changed the color of the spectrum trace to make it look like the line display of a spectrum analyzer and I turned off the peak hold display. I also changed the 'Waterfall Size (Expand Screen)' setting which lets you set the ratio of the waterfall to panadapter spectrum, displayed on the expanded spectrum scope display. The full table of settings is in the SCOPE Spectrum Scope EXPD/SET section on page 100.

SETTING THE SPECTRUM VIEW TO A 'LINE' VIEW

I don't like the default spectrum display, which is a filled spectrum with peak hold. It's a personal preference. You might like it. But I don't. I guess that I am used to spectrum analyzers and SDR receivers which have a line display rather than a filled-in trace. Although there is no option for a line display there is a 'workaround' that works for me.

If the Soft Key icons are not currently displayed at the bottom of the touch screen, hold <M.SCOPE> down for one second to enable the expanded screen.

Touch and hold <EXPD/SET> to open the 'Scope Set' menu. On the SCOPE SET menu, make the following changes.

<Max Hold>	OFF
<Waveform type>	Fill+Line
<Waveform Color (Current)>	R:30 G:30 B:191
<Waveform Color (Line)>	R:200 G:200 B:200

If you don't like the changes you can set the display lines back to default by touch and holding the menu item and selecting the 'Default' option.

MEMORY MANAGER

If you touch the text that says VFO A, VFO B or MEMO, or the memory number below that, you will open the VFO/MEMORY sub-menu.

This screen offers an alternative way to manage the radio memories compared to using the <MENU> <MEMORY> sub-menu.

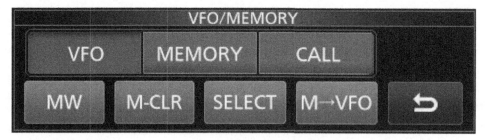

VFO sets the VFO mode and MEMORY sets the memory mode, in exactly the same way as pressing the V/M button.

In the VFO mode, MW saves the current VFO frequency, mode, and filter settings, etc. to the current memory slot as indicated below the VFO A or B indicator. M-CLR clears the current memory slot.

SELECT does not, as you might expect, select the memory and pop it into the VFO. Instead, if the memory mode is selected and the current memory slot is not blank, you can use SELECT to change the scan group for the currently selected memory channel. "Yay."

M->VFO loads the VFO with the information stored in the currently selected memory slot. You don't need to do this to use the selected frequency.

FILTER (FIL)

The current filter (1, 2, or 3) is indicated by the filter icon to the right of the blue mode indicator. A tiny white dot to the right of the icon indicates that Twin PBT (twin passband tuning) is active. Touch FIL repeatedly to cycle through the Filter 1, 2, and 3 bandwidths for the currently selected radio mode. A small popup appears briefly to show you the filter that you have selected.

Touch and hold the FIL icon to open the filter setup screen. Here you can see the effects of any Twin PBT adjustments, and you can adjust the actual filter bandwidth for each of the three filter pre-sets. For bandwidths greater than 500 Hz you can select a sharp or soft roll-off shape.

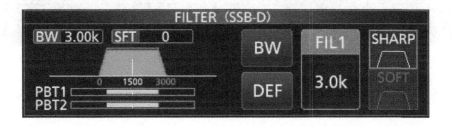

To adjust the bandwidth, select FIL1, FIL2, or FIL3 then touch the BW Soft Key. Adjust the bandwidth using the main VFO knob.

The FM and DV filter displays do not include PBT or adjustable filter shape and you can't change the bandwidths which are pre-set to 15 kHz, 10 kHz, and 7 kHz.

The AM and RTTY filter displays do not include the sharp and soft adjustable filter shape option.

MULTI SOFT KEYS

Pressing MULTI opens a set of Soft Keys on the right side of the display. The Soft Key icons are specific to the active band and the operating mode or function that you have selected. You can select the items by touching the appropriate icon and adjust them by turning the MULTI control knob. Touch and holding an icon usually allocates the adjustment to the MULTI knob and closes the sub-menu. See the full table of options on page 59.

SSB	SSB-D	CW	RTTY	AM FM DV

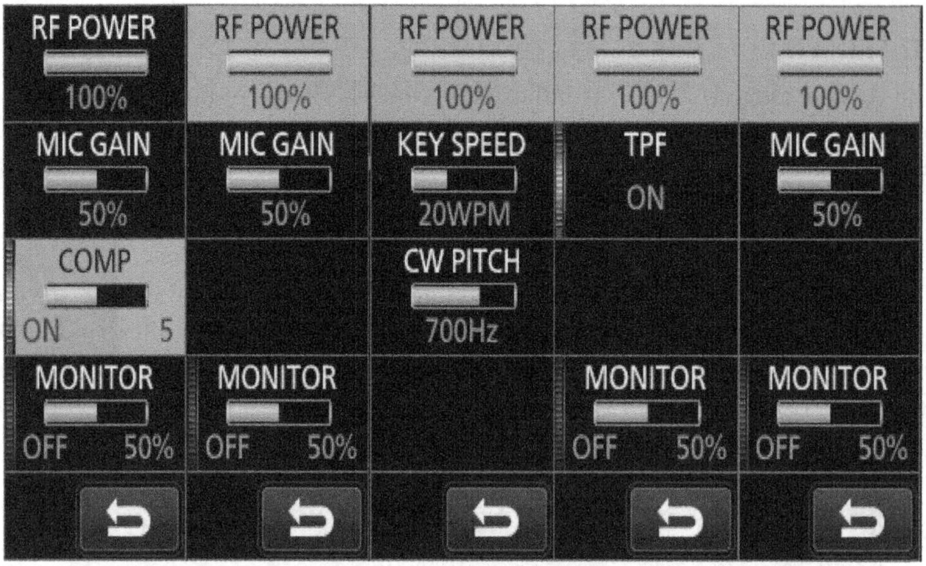

RF POWER
100%
RF POWER
100%
RF POWER
100%
RF POWER
100%
RF POWER
100%
MIC GAIN
50%
MIC GAIN
50%
KEY SPEED
20WPM
TPF
ON
MIC GAIN
50%
COMP
ON 5
CW PITCH
700Hz
MONITOR
OFF 50%
MONITOR
OFF 50%
MONITOR
OFF 50%
MONITOR
OFF 50%

The COMP, MONITOR, and TPF settings with 'on/off' features in addition to an adjustable control have a blue indicator at the left of the selection to indicate that the function is on.

Touching the screen anywhere outside of the menu area, touching the Return ⮌ icon, pressing MULTI again, or pressing EXIT will close the menu selection.

➢ RF POWER

Sets the transmitter's output power for the currently selected band. Press the MULTI button and touch the RF Power icon. Set the power for 100% by turning the Multi knob. 100% power is 100 Watts on 2 m, 75 Watts on 70 cm, and 10 Watts on 23 cm. You may need to reduce the transmitted power during satellite operation, particularly through linear transponders. Or when operating portable on battery power.

➢ MIC GAIN

The Mic Gain control is used to set the modulation level on voice modes. Turn the MULTI knob to change the setting. I adjust it on SSB mode with the compressor turned off. See 'Setting up the radio for SSB operation' on page 16. Once the Mic Gain has been adjusted correctly for the SSB mode it should be fine for FM, AM, and DV.

➢ COMP level

Touch to turn on the speech compressor. A blue indicator bar and the word ON indicate when the compressor is active. Also, COMP is displayed on the top line of the display. The default level is 5 and that level works for me. Turn the MULTI knob to change the setting.

➢ MONITOR

The Monitor Soft Key turns on the audio monitor so that you can hear the signal that you are transmitting. Turning the Multi knob adjusts the level of the monitor signal. 'Monitor' is not available in the CW mode because sidetone is always on.

A blue indicator and the word ON indicate that the transmit monitor is turned on.

Unless you specifically want to listen to your transmit signal it is less distracting to leave the Monitor turned off. The SSB voice keyer will be heard if the 'Auto Monitor' setting is ON, irrespective of the Monitor setting.

➢ KEY SPEED

When the radio is in CW mode the CW speed is adjustable from 6 wpm to 48 wpm. Turn the MULTI knob to change the setting.

➢ CW PITCH

The CW pitch is adjustable from 300 Hz to 900 Hz. It is a personal preference. I use 700 Hz. Turn the MULTI knob to change the setting.

➢ TPF (Twin Peak Filter)

The TPF (Twin Peak Filter) filter is available in the RTTY mode. It is very effective at lifting weak RTTY signals out of the noise. When TPF is turned on, a blue bar is displayed to the left of the control.

The TPF has a peak at the RTTY Mark and Space audio frequencies, emphasizing the RTYY signal and reducing noise elsewhere in the audio spectrum. TPF can make tuning an RTTTY signal a bit tricky, so it may be easier to turn it on after you have tuned to the RTTY signal.

Some of the MULTI menus are activated by holding down front panel buttons or Soft Keys on the FUNCTION display.

CW Break-in	VOX	Manual Notch	Noise Reduction	Noise Blanker
BKIN DELAY 7.5d	VOX GAIN 50%	NOTCH POSITION	NR LEVEL 4	NB LEVEL 50%
	ANTI VOX 50%	WIDTH MID		DEPTH 8
	DELAY 0.2s			WIDTH 50
	VOICE DELAY OFF			
↵	↵	↵	↵	↵

> CW Break-in delay

The CW Break-in delay submenu is activated in CW mode by pressing the FUNCTION button and then touch and holding the BKIN Soft Key or by holding down the VOX/BK-IN button. The delay setting affects the semi Break-in 'BKIN' mode. In semi break-in mode the Morse key or paddle will key the transmitter while the CW is being sent and the radio will return to receive after a delay when the key is released. The default delay is 7.5 dits at the selected key speed. In the full break-in mode, the transmitter returns to receive immediately after the key is released.

> VOX settings

The VOX submenu is activated in SSB, AM, or FM modes by pressing the FUNCTION button and then touch and holding the VOX Soft Key. Or by holding down the VOX/BK-IN button. The settings for VOX Gain, Anti-VOX, Delay, and Voice Delay are covered in the FUNCTION menu section, on page 144.

> Manual Notch

The Notch menu is accessed by holding down the NOTCH button on the front panel or by touch and holding the NOTCH Soft Key on the FUNCTION menu. The manual notch is very effective at removing annoying interference. The second icon allows you to set a wide, medium, or narrow notch. Start with narrow and if you still hear interference consider using the wider options.

You can see the effect of the notch filters by opening the Audio Scope. Select <MENU> <AUDIO>. Tune to a frequency where you can hear a carrier 'birdie.' You must have the receiver squelch open to see signals on the audio spectrum scope. Turn on the auto notch (AN) and you will see the carrier signal disappear from the audio spectrum display and you won't be able to hear it anymore. Enable the manual notch (MN) and you will be able to see a black zone on the audio spectrum indicating a deep null in the signal. Changing from narrow to mid or wide makes the null zone wider. Adjusting the manual notch position moves the nulled band across the audio spectrum.

There are two menu settings that can affect Notch operation. For AM and SSB you can select auto notch only, manual notch only, or the default choice of both auto and manual. Since the auto notch is best for interfering carriers, and the manual option allows you to place the notch where you want it on the audio passband, I can't imagine why you wouldn't want both options. But you can change it if you want to. <MENU> <SET> <Function> <[NOTCH] Switch (SSB)> or <[NOTCH] Switch (AM)>.

➢ NR level

The default noise reduction level is 5. Adjust the level to a point where the noise reduction is effective without affecting the wanted signal quality.

➢ NB level, depth, and width

The noise blanker LEVEL control (default 50%) sets the audio level that the filter uses as a threshold. Most DSP noise blankers work by eliminating or modifying noise peaks that are above the average received signal level. They usually have no effect on noise pulses that are below the average speech level. Setting the NB level below the average speech level will stop the noise blanker working. Setting it to an aggressive level may affect audio quality by clipping the audio.

The DEPTH control (default 8) sets how much the noise pulse will be attenuated. Too high a setting could cause speech before or after the interference pulse to be attenuated. This can cause the audio to sound 'choppy.'

The WIDTH control (default 50%) sets how long after the start of the pulse the output signal will remain attenuated. Set it to the minimum setting that adequately removes the interference. Very sharp short duration spikes will need less time than longer or multiple noise spikes such as lightning crashes.

Operating the radio

The Icom IC-9700 is a joy to use. I really like the feel of the main VFO tuning knob. The drag is adjustable, but I like it just the way it came from the factory. Generally, the ergonomics are good although some things take a bit of getting used to. For example, if you are using the Decode, Voice, or Keyer screens and you change mode, the screen does not follow the mode change.

VOICE KEYER

The Voice message option is available when the active VFO is in a voice mode or a voice-data mode. Press <MENU> VOICE> and touch a pre-recorded message T1 to T8 to send the voice messages. Recording and setting up the voice messages is covered on page 103.

Touch and holding a voice message key, T1 to T8 will cause the message to repeat until you touch the T message key again or press the PTT on the microphone.

OPERATING SPLIT

Working in the 'Split' mode is a very common requirement if you are trying to work a DX station on SSB or CW that has a 'pileup' of stations calling them. You cannot select split when the radio is in MEMO mode. It must be in VFO mode. You can't store a 'split' pair of frequencies into a memory slot either.

Pressing SPLIT engages split mode. Split operation in the IC-9700 is different from many transceivers where a pre-defined 5 kHz or similar offset is applied. In the IC-9700 using split means that the transceiver will receive on the frequency indicated by the 'A' VFO and transmit on the frequency indicated by the 'B' VFO, (or vice versa). The second VFO might be on a completely different part of the band and using a different mode. So, you do have to be careful.

To work 'Split' you set the current VFO, we will assume that is VFO A, to receive the frequency that the DX station is using. You will transmit on the frequency indicated by VFO B. Initially you should press and hold the SPLIT button to enable split operation. This will set the transmit VFO to the offset stored in the menu setting. After the split offset has been set you can turn split on or off by pressing the SPLIT button. Split operation is indicated with an orange 'Split' indication. VFO A and B can be reversed using the A/B button. It makes no difference which you use for transmitting.

Holding SPLIT down can have two different effects depending on the split menu setting. If QUICK SPLIT has been set to ON, holding down the SPLIT button sets the split nominated in the Split Offset menu setting. If QUICK SPLIT has been set to OFF, holding down the SPLIT button has no effect.

➢ Split menu settings

To change the split menu settings, select; <MENU> <SET> <Function> <SPLIT>.

- QUICK SPLIT should be set to ON. Otherwise, pressing and holding the SPLIT button has no effect.

- SPLIT OFFSET can be set to your favorite split offset in MHz. For example, +0.005 MHz for SSB.

- SPLIT LOCK only has an effect if you are using Dial Lock.

    o If you have Dial Lock turned on and Split Lock turned off, you will not be able to change the receiver frequency or the transmit frequency while holding the XFC button.

    o If you have Dial Lock turned on and Split Lock turned on, you will not be able to change the receiver frequency, but you will be able to change the transmit frequency while holding the XFC button.

➢ Manually setting the split offset

There are a couple of ways to set the split offset while operating. I prefer method 1.

Method 1:

- Press and hold SPLIT or A/B to set the transmitter VFOs to the stored split offset.

- Press the XFC button and check that both VFOs are set for the same mode (i.e. CW or SSB)

- Hold down the XFC button (above the RX/TX LED) and use the main VFO knob to set the frequency that you want to transmit on. The split offset is displayed near the Hz digits of the VFO. This method has the advantage that you can hear the frequency that you will transmit on while the XFC button is held down.

Method 2:

- Press and hold SPLIT or A/B to set the transmitter VFOs to the stored split offset.

- Use the XFC button to make sure both VFOs are set for the same mode (i.e. CW or SSB)

- Touch the MHz digits of the VFO display.

- Touch F-INP

- Enter the split in kHz. For example, for a plus 5 kHz split touch <5> then touch the SPLIT Soft Key on the F-INP screen. For a -2 kHz split touch <(-)> <2> then touch the -SPLIT Soft Key on the F-INP screen

OPERATING CW MODE

In CW mode the receiver should be tuned to the exact frequency of the CW station. If you have the spectrum scope turned on, the centerline or green receive marker should be exactly aligned with the signal shown on the spectrum and waterfall display. At that point, the audio tone will be very close to the audio tone frequency set by the CW PITCH control. Press AUTO TUNE while receiving the CW signal to net the receiver frequency exactly.

➢ Break-in setting

In CW mode the VOX/BK-IN button controls the break-in settings. They can also be changed by pressing the FUNCTION button and using the BKIN Soft Key icon.

The transceiver will not automatically transmit CW unless either semi break-in (BKIN) or full break-in (F-BKIN) has been selected. The current break-in setting is displayed at the top of the screen. The break-in setting affects the sending of keying macro messages and Morse Code sent from a key or paddle, but not CW sent from an external computer program.

- Full break-in mode F-BKIN will key the transmitter while the CW is being sent and will return to receive as soon as the key is released. This allows for 'QSK' reception between CW characters.

- Semi break-in mode BKIN will key the transmitter while the CW is being sent and will return to receive after a delay when the key is released. Touch and hold the BKIN Soft Key or hold the VOX/BK-IN button to adjust the delay. Turn the Multi knob to change the setting. The default is a period of 7.5 dits at the selected key speed.

- With BKIN OFF you can practice CW by listening to the side-tone without transmitting. The transceiver can be made to transmit by pressing the TRANSMIT button, pressing the PTT button on the microphone, sending a CI-V command, or grounding the SEND line on the ACC jack.

I suggest that you don't use full break-in on the 2m band. The noise from the PTT relay will drive you nuts. On the 70 cm band, the keying is done with silent Pin Diode switches and on the 23 cm band, the micro-relay used for PTT is very quiet.

➤ MONI (Monitor)

The MONI (transmit monitor) function is disabled in CW mode because the sidetone is always turned on. If you don't want sidetone you can turn down the level to zero. (See Sidetone below).

➤ Sidetone

The CW sidetone is 'always on' but you can set the level to zero if it is annoying, or your key, paddle or bug has its own sidetone.
In CW mode select <MENU> <KEYER> <EDIT/SET> <CW-KEY SET> <Sidetone Level>

➤ Auto Tune

The AUTO TUNE button pulls the receiver onto the correct frequency while receiving a CW signal. It adjusts the receive frequency until the received CW signal is at the tone set by the CW PITCH control. At that point, the received CW signal is exactly netted with the transmit frequency. You may have to press the button a couple of times to get the tuning exact.

➤ Key speed

While in CW mode press the MULTI button and touch the KEY SPEED icon. Turning the MULTI knob sets the speed of the electronic keyer, including CW sent from the KEYER SEND macros. The key speed is adjustable from 6 wpm to 48 wpm.

➤ CW pitch

The CW PITCH control on the MULTI menu changes the pitch of a received CW signal without changing the receiver frequency. You can set the control so that CW sounds right to you.

The CW PITCH frequency is depicted below the center of the filter passband image. Touch and hold the FIL icon at the top of the screen. The center frequency of the passband is the pitch frequency.

➤ CW message keyer

When in CW mode press MENU then KEYER to show the eight CW messages. The CW messages are handy for DX or Contest operation or just to save you sending the same message over and over. They are great for sending CQ on a quiet band.

Touching one of the **M1 to M8 Soft Keys** sends the CW message. You can stop it by touching the Soft Key again or by sending with the key or paddle

Touch and hold one of the **M1 to M8 Soft Keys** to keep sending the message repeatedly until you stop it by touching the Soft Key again or by sending with the key or paddle.

The M2 Soft Key is setup for contesting with an incrementing number after the signal report. If you need to send the same number again, you can decrement the number with the -1 Soft Key. The radio remembers the last number sent even if you have turned it off in the meantime, so you won't lose the number during a 48 hour contest. See 'KEYER sub-screen Soft Keys' on page 106 to set up the keyer memories.

➢   ¼ tuning speed

The digital modes and CW allow the use of the ¼ tuning function. Select <FUNCTION> <1/4> to turn the function on or off. The function slows down the tuning rate of the VFO to make tuning in narrow CW and digital mode signals easier. It is indicated with a ¼ icon to the right of the 10 Hz digit of the frequency display. It is indicated above the 1 Hz digit if the 1 Hz tuning step is active. Touch and hold the 10 Hz digit on the VFO display to switch to the 1 Hz tuning step.

OPERATING RTTY

The radio supports three kinds of RTTY operation. Firstly, there is the onboard RTTY decoder. Which can be used with the eight RTTY message memories.

In this mode, you can take advantage of the excellent TPF (twin passband filter). I recommend using the TPF filter all the time because it really helps with accurate decodes. The second method is to use external PC software such as; MixW, MMTTY, MMVARI, Fldigi, etc. with AFSK (audio frequency shift keying). AFSK uses two audio frequencies to create the frequency-shift keying in the SSB mode. The third method is the FSK mode which uses a digital signal to key the transceiver to predefined mark and space offsets.

➢   RTTY decode

Set RTTY mode with a narrow panadapter ±2.5 kHz or ±5 kHz. This must be done **before** you turn on the DECODE screen.

Selecting <MENU> <Decode> while in RTTY mode displays the internal RTTY decoder. The DECODE option is not visible on the main MENU unless the radio is in the RTTY mode. Selecting DECODE will shrink the panadapter display.

Tune the VFO frequency so that the panadapter center line or receive marker is aligned with the right (higher frequency) of the two RTTY lines.

Then fine-tune until the two RTTY signal peaks are lined up with the vertical lines on the audio spectrum display on the DECODE screen. This is easier to do with the TPF filter off. There is also a tuning indicator at the top left of the audio spectrum display. After you have the signal tuned you can turn on the excellent TPF (twin peak filter) which is specially designed to maximize the two RTTY tone frequencies. The TPF can be turned on or off using <MULTI> <TPF>.

If you touch the TX MEM Soft Key, the RTTY message keys RT1 to RT8 are displayed. They can be triggered by touching the relevant Soft Key or from a keyboard or button pad attached to the microphone connector. After the message has been selected, the display reverts to the decode screen.

The 'Decode' screen can be made larger by touching the EXPD/SET Soft Key. This means that more decoded text can be displayed, but it overlays the panadapter display.

➢ **TPF filter**

Don't forget to use the fabulous TPF (Twin Peak Filter) filter it is very effective at lifting weak RTTY signals out of the noise. Tune in the signal first then turn on the filter by pressing the MULTI button and touching the TPF icon to enable the filter. When TPF is on, a blue bar is displayed to the left of the control.

➢   ¼ tuning speed

The RTTY mode allows the use of the ¼ tuning function. Select <FUNCTION> <1/4> to turn the function on or off. The function slows down the tuning rate of the VFO to make tuning in narrow digital mode signals easier. It is indicated with a ¼ icon to the right of the 10 Hz digit of the frequency display. It is indicated above the 1 Hz digit if the 1 Hz tuning step is active.

➢   AFSK RTTY from an external PC program

To use an external PC based digital modes program for transmitting AFSK RTTY you must use the USB-D DATA mode, not the RTTY mode. First, ensure that the active VFO is in SSB mode and then touch <DATA>. The mode should change to USB-D. You can transmit RTTY on LSB-D but it is not standard even on low bands.

*Audio is sent to the PC when it is in any mode, so you can use your digital mode PC software to see and decode RTTY signals. But you should be in the USB-D DATA mode to transmit accurate AFSK RTTY from your digital mode PC software. In fact, with the default 'Data Off Mod' setting, you **must** be in the USB-D DATA mode to transmit any audio from your digital mode PC software.*

If you prefer to use your favorite external digital mode program to send RTTY, the easiest method is to use AFSK rather than FSK. In AFSK the RTTY signal is sent to the transceiver as audio tones rather than a digital signal. I use MixW because I like the log function and it seems to decode well. But any digital mode software should be OK. I have found that early versions of MixW will not communicate with the radio due to the Icom addressing on the USB COM port. But the latest MixW V3.1.1 and V4 versions are fine.

➢   RF Power in RTTY mode

You can run 100 Watts, but if you are prone to very long 'overs' I suggest derating the power to 75 Watts. Press MULTI select the RF POWER icon and reduce RF power to 75% by turning the Multi knob. Keep an eye on the temperature meter on the multi-function meter display. If the transceiver is running hot, de-rate the transmitter power.

➢   FSK RTTY from an external PC program

An advantage of using FSK rather than AFSK from an external PC program is that you use the RTTY mode on the radio rather than the USB-D data mode. That means that you can use the TX MEM messages and the TPF (twin peak filter).

Select the RTTY mode. There are no levels to set in the FSK mode. There are instructions for setting up MMTTY for FSK operation on page 38. I have not been able to get MixW to send FSK RTTY, but the AFSK mode works fine.

OPERATING USING THE SCAN MODE

The front panel SCAN button has slightly different functions to the main menu scan mode including some neat memory slot scanning options. See SCAN button on page 64.

Selecting <MENU> <SCAN> displays the SCAN Menu. Touching the screen, PTT from the Mic, or touching a Soft Key again will halt a scan that is in progress.

The radio has three scan modes. You can scan a range either side of the current VFO setting, between pre-set frequencies, or you can scan through all or some of the memory channels.

- The ◢F Soft Key starts a scan from just below the current VFO frequency to just above it. You can set scan ranges from ±5 kHz to ±1 MHz using the ◢F SPAN Soft Key. The current fast or normal tuning step is used.

- Touch PROG to scan between pre-set frequencies. PROGRAM SCAN flashes while the scan is progressing and the decimal points in the frequency display flash. Touch and hold PROG to select one of the three Program Scan ranges. They are stored in the main memory bank as 1A – 1B, 2A – 2B, and 3A to 3B.

- MEMO replaces PROG if the radio is in Memory mode instead of VFO mode. It is used to scan through selected memories from the 99 stored memories. If you touch SELECT during a memory scan, the scan will restrict itself to the nominated span group. You can change the span group using the SEL No. Soft Key, but only while a scan is in progress.

Set the squelch so that the radio is squelched before using the Scan mode and the Scan will stop when a signal is encountered. You can start the scan again by touching the PROG or ▲F Soft Key. Alternatively, if Scan Resume is set to 'ON' the scan will resume after a time. If you touch the FINE Soft Key **after** a scan has started, the scan will slow down but not stop when the squelch opens. After it passes through the signal and the squelch closes again, the scan speeds up again. I like the ▲F scan mode with 'Scan Resume' set to ON.

TIP: Create a set of panadapter band edges with the same bandwidth (up to 1 MHz) as the program scan 1A – 1B, 2A – 2B, or 3A to 3B range. Then you can use the FIX panadapter mode to see the entire scan range as a scan is in progress.

In effect, there are four scan speeds. The scan uses the VFO tuning step and the SET menu lets you select either a Fast or Slow scan speed. The fastest scan is when the VFO is in the 1 kHz step mode and the scan speed is set to fast. The slowest scan speed is when the VFO is set to 1 Hz steps and scan speed is set to slow.

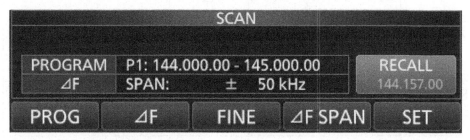

Scan controls in the VFO mode

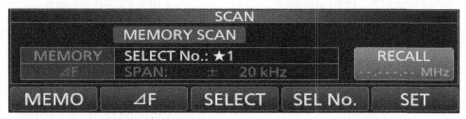

Scan controls in MEMO mode while scanning

Soft Key	Function	Hold	Menu Setting / Notes
PROG	Touch PROG to scan between the pre-set frequencies. PROGRAM SCAN flashes while the scan is progressing and the decimal points in the frequency display flash.	Select P1, P2 or P3 pre-set ranges	The P1, P2, and P3 ranges are stored in the main memory bank as 1A – 1B, 2A – 2B, and 3A to 3B. <MENU> <MEMORY>
◢F	Span frequencies below and above the current VFO setting. ◢F SCAN flashes while the scan is progressing and the decimal points in the frequency display flash.	None	The range of the span is set by repeatedly touching the ◢F SPAN control.
FINE	Touch FINE while a scan is in progress and the scan rate will slow but not pause when a signal opens the receiver squelch.		
◢F SPAN	Sets the range of the ◢F scan.	None	±5 kHz, ±10 kHz, ±20 kHz, ±50 kHz, ±100 kHz, ±500 kHz, ±1 MHz.
RECALL	None	VFO recall function	Touch and hold to reset the VFO to the original frequency.
SET	Sub-menu to change Scan speed and Scan resume function	None	Scan speed, resume, pause timer, resume timer, temporary skip, main dial
MEMO	MEMO replaces PROG if the radio is in Memory mode. Scan all or selected memory channels.	None	Memo steps through all the stored memory slots, or through channels that are tagged with a scan group tag, *1, *2, *3, *123.

| SELECT | When not scanning, SELECT changes the scan group of the current memory channel. While scanning SELECT selects all channel or scan group only scan. This is buggy and sometimes it stops the scan instead. Using SEL no. then SELECT seems to fix the issue. | Clears one or all scan groups | Touch and hold clears one or all scan group settings. Be careful. |
| SEL No. | Selects the group of memory channels that will be scanned. *1, *2, *3, or *123. | None | SEL No. is only displayed in memory scan mode and only while the scan is running. |

OPERATING EXTERNAL DIGITAL MODE SOFTWARE

Once the connection has been established between your digital mode software and the radio, the operation of the radio is mostly controlled from the external software. The main thing to remember is that for most modes you must be in the Data mode (USB-D or LSB-D) to transmit from an external program. In the other modes, the transmitter will key but the modulation will not be transmitted. The exceptions are DTR keyed modes. CW will work in the CW mode and FSK (not AFSK) RTTY will work in the RTTY mode.

*Note that audio is sent to the PC when it is in any mode, so you can use your digital mode PC software to see and decode PSK signals. But you **must** be in the USB-D DATA mode to transmit PSK from your digital mode PC software. You can spoof the default 'Data off Mod' setting by changing it to 'MIC, USB' which will allow audio from the USB cable to be transmitted in non-data modes, but the settings may be incorrect for digital modes, VOX and the speech processor may be activated, and if MIC is selected, the microphone will be live while transmitting.*

➢ WSJT-X for JT65 or FT8

If you are using WSKT-X for FT8 modes, the software 'Split operation' should be set to 'Rig.' See page 28. This allows the software to use 'Split' to change the transmit VFO frequency in order to reduce the possibility of audio harmonics affecting other FT8 users.

For example, if you are transmitting at 600 Hz above the nominated FT8 frequency there could be harmonics at 1200 Hz and 1800 Hz. The function changes the VFO on transmit so that the audio tone is higher, making any possible audio harmonics fall outside of the transmitter's transmit bandwidth.

SAVING A FREQUENCY TO A MEMORY SLOT

Method 1:

Touch the memory channel number displayed under the word MEMO or VFO on the right side of the touch screen display.

This will bring up the very handy VFO/MEMORY sub-menu screen. Touch the MEMORY Soft Key. Use the Up and Down buttons on the microphone to step through the saved memory channels or touch and hold the kHz M-CH button to activate the M-CH memory channel step mode and use the MULTI knob. You can also touch CALL to select one of the two call channels. Touching VFO returns the radio to VFO mode.

In VFO mode touch and hold the MW Soft Key to write the current frequency mode to the currently selected memory slot. This will overwrite the previous contents. Note that there is no way to add a name to the memory slot. You will have to edit it using the main memory manager. (See method 2).

In MEMORY mode you can touch and hold the M-CLR Soft Key to delete the contents of the currently selected memory slot. But, you do not have to do this to overwrite the contents.

SELECT does not do what you might expect. In MEMORY mode touch the SELECT Soft Key to select place the memory slot into one of the three scan groups. Touch and hold, allows you to clear a scan group – use with care!

Touch and hold the M->VFO Soft Key to copy the current memory settings into the active VFO.

Method 2:

Normally you will be in the VFO mode with the radio settings ready to be saved into a memory slot. If you are in MEMO mode, you can duplicate or edit an already stored memory channel.

Select <MENU> <MEMORY> to open the memory manager. You will see a table that contains the 99 available memory slots for the current band.

It also has slots labeled 1A, 1B, 2A, 2B, 3A, and 3B which set the scan limits for the scan mode.

The C1 and C2 slots are for calling channels. Put your most commonly used repeater or simplex channels into a calling channel slot. The one that the radio is left tuned to while you are pottering around in the shack.

Use the MULTI knob or the UP DOWN Soft Keys to select an unused memory slot, (or a slot that you wish to overwrite).

This is a bit tricky! Touch the three-line icon to the right of the blank cell, or touch and hold anywhere else in the blank cell. It is best to get used to touching the icon.

If the cell is a previously unused slot there will be only one option. 'Memory Write.' Touch <Memory Write> and then <YES>. This should save the current frequency, duplex shift, and tone settings. But it does not store a name for the channel.

If the cell is a currently used slot there will be three options. 'Edit Name', 'Memory Write' and 'Memory Clear.' It is not necessary to clear a memory slot before overwriting it. Touch <Memory Write> and then <YES>. This should save the current frequency, duplex shift, and tone settings into the selected memory slot.

Touch the three-line icon to the right of the memory slot and this time select the <Edit Name> option. Enter the name for the stored frequency or channel and then touch the ENT icon to exit.

RECALLING A STORED MEMORY CHANNEL

Method 1:

Press V/M to change to the MEMO memory channel mode and step through the channels using the MULTI knob or the Up and Down buttons on the microphone.

You may have to select the M-CH memory channel step mode by touch and holding the kHz M-CH button.

Method 2:

Touch the memory channel number displayed under the word MEMO or VFO on the right side of the touch screen display.

This will bring up the very handy VFO/MEMORY sub-menu screen. Touch the MEMORY Soft Key. Use the Up and Down buttons on the microphone to step through the saved memory channels or touch and hold the kHz M-CH button to activate the M-CH memory channel step mode and use the MULTI knob. You can also touch CALL to select one of the two call channels. Touching VFO returns the radio to VFO mode.

You can touch and hold the M->VFO Soft Key to copy the current memory settings into the active VFO, but you don't have to do this to use the stored channel.

The VFO/MEMORY sub-menu

Method 3:

Press V/M to change to the MEMO memory channel mode, then select <MENU> <MEMORY> and use the UP DOWN Soft Keys or the MULTI knob to select a channel.

Note: If you select a memory channel while in VFO mode instead of MEMO mode, the VFO will not change to the channel that you have selected. However, when you do change to MEMO mode using the V/M button the channel will be selected.

The main menu, memory manager

Menu

MENU is the first of the buttons located below the touch screen. Initially, you will use the MENU button a lot as you set up the radio's many different options and features. Some of the displayed icons change according to the mode that the radio is in. For example, if the radio is set for SSB operation there is a VOICE option. In CW mode this is replaced by KEYER. In RTTY mode it becomes DECODE.

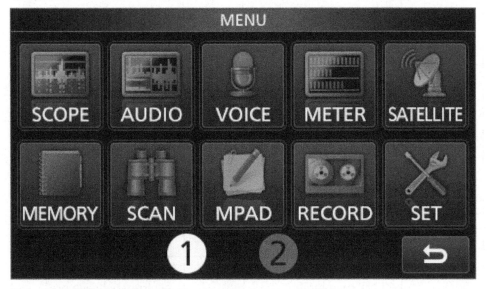

The main MENU in SSB mode (page 1 of 2).

SCOPE

Touching the SCOPE Soft Key turns on the panadapter spectrum and waterfall display with the SCOPE Soft Key icons at the bottom. The size of the panadapter is determined by the EXPD Soft Key setting.

You can also activate the panadapter spectrum and waterfall display by press and holding the M.SCOPE button.

Pressing the M.SCOPE button (rather than press and holding it) cycles through two displays. One with a small panadapter with no Soft Key controls and the other with large VFO numbers and no panadapter. But for the following discussion, we need the display option that includes the Scope Soft Keys.

➢ SCOPE Spectrum Scope Soft Keys

Select <MENU> <SCOPE> or hold down the M.SCOPE button.

| < 1 > | EDGE | HOLD | CENT/FIX | EXPD/SET |

| < 1 > | SPAN | HOLD | CENT/FIX | EXPD/SET |

	Function	Hold	Menu Setting / Notes
< 1 >	Change to Menu 2 selections.	None	
EDGE	In the fixed (FIX) spectrum mode, EDGE cycles through three pre-set band edges per frequency zone.	None	To set the band edges <M.SCOPE> <hold EXPD/SET> <Fixed Edges>
SPAN	In the center (CENT) spectrum mode, Span cycles through the Span settings.	Reset to ±2.5 kHz	±2.5 kHz, ±5 kHz, ±10 kHz, ±25 kHz, ±50 kHz, ±100 kHz, ±250 kHz, ±500 kHz
HOLD	Freezes the spectrum scope and waterfall.	None	
CENT/FIX	Toggles the FIX and CENT span modes. SCROLL-F displays the spectrum with a span defined by the pre-defined band edges. SCROLL-C displays the spectrum with a span defined by the SCAN bandwidth.	None	FIX mode displays the spectrum between pre-defined band edges. CENT displays the spectrum below and above the active VFO frequency.
EXPD/SET	Touch EXPD/SET to turn on or off the expanded spectrum scope display. Touch and hold EXPD/SET to enter the screen setup menu.	Enter SET menu	See next table for the SET menu

| < 2 > | REF | SPEED | MARKER | EXPD/SET |

< 2 >	Change to Menu 1 selections.	None	
REF	Allows adjustment of the Spectrum display reference level.	None	Use the Main VFO knob to adjust REF level. Touch REF again to return to Menu 2. Touch and hold DEF to set REF to 0 dB.
SPEED	Cycle through FAST, MID, SLOW panadapter and waterfall speeds.	None	Indicated with blue >>> icons at the top of the Spectrum Scope
MARKER	Displays the orange (T) transmit frequency marker. In the FIX spectrum mode, the green (R) receive frequency marker is always displayed.		>> at the right side of the panadapter or << at the left side of the panadapter indicates that the marker is lower or higher than the panadapter can show on its current setting.

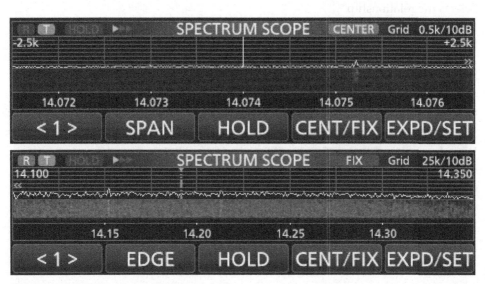

The Centre and Fixed panadapter modes

➢ SCOPE Spectrum Scope EXPD/SET settings

Select <MENU> <SCOPE> or hold the M.SCOPE button for one second, until the large or medium panadapter is displayed. Then hold the EXPD/SET Soft Key down for one second to display the 16 scope settings. Please read the important notes about the carrier point display following the table.

Setting	IC-9700 default	ZL3DW setting	My Setting
Scope during TX (CENTER) Sets the scope to display the transmitted signal while you are transmitting.	On	On	
Max Hold Sets the peak hold function on the spectrum display	10-second hold	Off	
CENTER Type Display Selects the relationship between the CENT scope line and the receiver filter	Filter Center [1]	Carrier Point Center (Abs Freq) [2 & 3]	
Marker Position (FIX Type) Selects the relationship between the FIX scope marker line and the receiver filter	Carrier Point [2]	Carrier Point [2]	
VBW (video bandwidth)	Narrow	Narrow	
Averaging smooths the spectrum display by averaging 2, 3, or 4 sweeps.	OFF	Averaged over 2 sweeps	
Waveform type. Filled or filled with a line. See 'Useful Tips' to emulate a line display.	Fill	Fill+Line (with color modification)	
Waveform Color (Current)	R:172 G:191 B:191	R:30 G:30 B:191	
Waveform Color (Line)	R:56 G:24 B:0	R:200 G:200 B:200	
Waveform Color (Max Hold)	R:45 G:86 B:115	R:45 G:86 B:115	

Waterfall display	ON	ON	
Waterfall speed	MID	MID	
Waterfall Size (Expand Screen)	MID	SMALL	
Waterfall Peak Color Level	Grid 8	Grid 8	
Waterfall Marker Auto-hide	ON	ON	
Fixed Edges	Displays three adjustable frequency zones for each of the three bands. See 'Setting up the panadapter band edges' on page 73.		

1. Filter center places the VFO frequency marker in the middle of the filter bandwidth, at the center of a USB or LSB signal. In my opinion, this is just plain wrong.

2. Carrier Point (Center) places the VFO frequency marker at the carrier point. The left side of an upper sideband (USB) signal or the right side of a lower sideband (LSB) signal. In FIX mode bottom of the display shows the actual frequency. In CENT mode, the bottom of the display shows the span in kHz.

3. Carrier Point Center (Abs Freq) places the VFO frequency marker at the carrier point. The left side of an upper sideband (USB) signal or the right side of a lower sideband (LSB) signal. The bottom of the display shows the actual frequency in MHz.

AUDIO

➤ AUDIO sub-screen Scope Soft Keys

Selecting <MENU> <AUDIO> displays the received audio level and spectrum on the AUDIO SCOPE. While transmitting it displays the transmit audio level and spectrum. The audio scope will not display received audio if the receiver is muted. You can see the effect of the filter passband on the width of the audio spectrum display and the effect of the NR noise reduction filter and the manual notch filter.

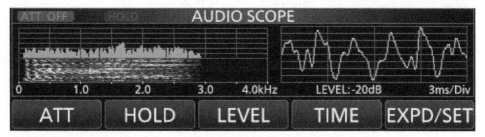

The audio scope works while receiving or transmitting

Soft Key	Function	Hold	Menu Setting / Notes
ATT	Adds attenuation to reduce the level of the audio spectrum and the brightness of the waterfall.	Resets to 0 dB.	Selectable 0dB, 10 dB, 20 dB, or 30 dB.
HOLD	Freezes the audio spectrum scope, the waterfall, and the oscilloscope.	None	
LEVEL	Changes the gain of the oscilloscope display.	None	Selectable 0dB, -10 dB, -20 dB, or -30 dB.
TIME	Changes the time-base of the oscilloscope display.	None	Selectable 1 ms, 3 ms, 10 ms, 30 ms, 100 ms, or 300 ms per division.
EXPD/SET	EXPD makes the audio spectrum and oscilloscope display taller.	SET mode see table below	See the Audio Scope SET mode table below

> AUDIO Scope SET settings

Select <MENU> <AUDIO> <SET> to display the Audio settings. You can set the spectrum, waterfall, and oscilloscope trace color, or turn the waterfall off.

Setting	IC-9700 default	ZL3DW setting	My Setting
FFT Scope Waveform Type	Fill	Line	
FFT Scope Waveform Color	R:51 G:153 B:255	R:51 G:153 B:255	
FFT Scope Waterfall Display	ON	ON	
Oscilloscope Waveform Color	R:0 G:255 B:0	R:0 G:255 B:0	

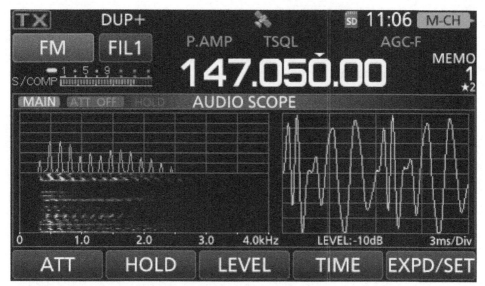

The audio scope expanded display showing a received signal

If both receivers are running, you can select a view of audio from the sub-receiver, by touching the orange MAIN (or SUB) Soft Key on the Audio Scope display.

VOICE

➤ VOICE sub-screen Soft Keys

The Voice option is only available when the active VFO is in a voice mode or a voice-data mode. Touch T1 to T8 to send the voice messages. You must have an SD card installed to use the voice keyer. Each of the eight 'voice messages' can be up to one and a half minutes long. Mine are all less than ten seconds long.

You can use four buttons connected via the microphone connector to send the first four messages. Set <MENU> <SET> <Connectors> <External Keypad> <VOICE>.

To record a voice message, select <MENU> <VOICE> <REC/SET> <REC>.

- Touch the message number T1 – T8 that you want to record. A recording screen will pop-up. The three buttons are Record, Playback, and Stop.

- You can record the message using your normal microphone. Do not press the PTT button. You don't need to transmit to make a recording. This is handy. I recorded a new message during a contest the other day.

  o You set the recording level by speaking into the microphone in the same way that you would when you are "on the air." Touch the MIC GAIN Soft Key at the right of the popup window.

You can adjust the level with the plus and minus buttons or with the main VFO knob. Set the level so that the peak level hold line just hits 100%. Don't worry if the occasional peak is higher than 100%. The transmitter limiter will deal with that. I found the default 50% level to be fine.

o   When you have set the Mic Level you can proceed with recording the messages. I write down exactly what I want to record. It really helps. Practice it a couple of times.

o   Touch the red 'record' icon and speak normally into the microphone. Don't press the PTT on the microphone. Touch the square 'stop' icon immediately after you stop speaking. It's best if you have your finger ready, poised over the button.

o   You can playback the recording by touching the triangle 'play' button. If you can't hear it, turn up the main receiver volume.

o   If you don't like the recording, just record a new message. It will overwrite the old message automatically.

o   When you are happy with the recording. Touch the return ⟳ icon and record the next message in the same way.

• You have to make the recording before adding a name to the T button. To add a name to the recording, touch <REC/SET> <REC> then hold the message number (T1 – T8). Then select <Edit Name>. Type the name of the message, for Example 'CQ short,' or 'my QTH.' The names are displayed on the eight 'voice message buttons.' They can be up to 16 characters long, but the button can only display about eight characters.

• Make sure to touch <ENT> or your changes will be lost when you exit the screen. To exit the recording screen, touch the return ⟳ icon or press the EXIT button.

Soft Key	Function	Hold	Menu Setting / Notes
T1 to T8	T1 to T8 sends the voice message.	Repeat	
REC	Record voice messages (see above)	None	<REC/SET> <REC>
Auto Monitor	'ON' lets you hear the voice message on playback	None	<REC/SET> <SET> <Auto Monitor>

Repeat Time	Set the time between message repeats. The default is 5 seconds.	None	<REC/SET> <SET> <Repeat Time>
TX LEVEL	Adjust TX LEVEL while the message is playing to get the same RF power on SSB as you would if you were speaking into the microphone. i.e. peaking to 100%.	None	Use the main VFO knob to adjust. The default level is 50%.

➤ Setting the level of voice keyer messages

To set the VOICE TX voice keyer transmit level.

a. Turn off the second receiver. Select a vacant frequency, (not a repeater input or output). Set SSB mode. Bring up the VOICE TX screen by pressing <MENU> <VOICE>

b. Touch the TX LEVEL Icon and then trigger one of the voice messages. Use the MAIN VFO knob to set the transmit level. Set the level so that you get peaks to 100% transmitter power on SSB. Try sending the same message by talking into the microphone. The aim is to have the recording sound exactly like you are using the microphone. I ended up leaving the TX LEVEL at 50%.

c. You can check that the level is about right by selecting the ALC meter by touching the meter scale until the right meter appears. The ALC reading should be similar when keying a voice message to when you are speaking into the microphone. Similarly, if you are using the compressor, you can select the COMP meter and compare the amount of compression being developed. Reduce or increase the TX LEVEL until the readings are similar or a little less than the microphone levels. Do not adjust the COMP level or the MIC GAIN or you will end up going in circles.

d. If you have set the Auto Monitor setting to ON (default) you will be able to hear the message as it is transmitted. <MENU> <VOICE> <REC/SET> <SET> <Auto Monitor> <ON>.

e. Touch and holding any voice keyer message icon causes the message to repeat at intervals set by the Repeat Time function. <MENU> <VOICE> <REC/SET> <SET> <Repeat Time>.

KEYER

The CW Keyer menu option is visible if the radio is in CW mode. Select <MENU> <KEYER>. It is used to set all the CW functions and to send the eight CW messages.

➢ KEYER sub-screen Soft Keys

Touching one of the **M1 to M8 Soft Keys** sends the CW message.

Touch and hold one of the **M1 to M8 Soft Keys** to keep sending the message until you stop it by touching the Soft Key again or by sending with the key or paddle.

You can use four buttons connected via the microphone connector to send the first four messages. Set <MENU> <SET> <Connectors> <External Keypad> <KEYER>.

Although there is a counter to increment the contest report number, there is no way to enter a callsign into a QSO or contest exchange other than editing the message for every QSO. You could send the callsign manually and then use the keyer to send the rest of the message. For contests, it's probably easier to use N1MM Logger +.

The -1 Soft Key: If you get a busted contest QSO you can decrement the counter by touching -1 on the Keyer message screen. The next M2 message will repeat the previously sent number. You can't manually increment the counter, but you can set it to any number you like using <MENU> <KEYER> <EDIT/SET> <001 SET> <Present Number>.

The CW message keyer screen

To **edit a CW message**, select <MENU> <KEYER> <EDIT/SET> <EDIT>.

- Touch the message that you want to edit and then touch the <EDIT> icon. A keyboard will appear on the touch screen.

- CW messages can be up to 70 characters long.

- A Caret ^ symbol removes the space between two letters, for example, ^AR.

- Press ENT to save your changes.

- Message M2 has an upward pointing arrow. It signifies that this message slot has an auto-incrementing counter used for contest exchanges.

  In operation, the * is replaced with the contest number. The default message for the M2 message is UR 5NN * BK, which will send UR 5NN 001 BK.

  Only one message can send the contest number. You can move the contest number (*) to another message but you must delete the * in the M2 message first. Also, if you do change the message that has the contest number (*), you have to tell the radio which message now has the * by changing the trigger. <EDIT/SET> <001 SET> <Count Up Trigger>.

  If you don't work contests you can delete the * and use the M2 message slot for a different message.

- When you have finished editing all the messages touch the return↺ icon to exit.

001 SET: On the <MENU> <KEYER> <EDIT/SET> <001 SET> menu you can set

a) The **number style** used for automatic contest numbers.

  a. Normal numbers, 001, etc.

  b. ANO style A=1, N=9, O=0

  c. ANT style A=1, N=9, T=0

  d. NO style N=9, O=0

  e. NT style N=9, T=0

b) **Count Up Trigger**. Sets which macro is using the automatic 'Count Up' feature. The default is M2. It must be the macro that has the contest number star (*) in the text.

c) The **Present Number**, usually 001. Note that if you get a busted contest QSO you can decrement the counter by touching -1 on the Keyer message screen.

CW-KEY SET: On the <MENU> <KEYER> <EDIT/SET> <CW-KEY SET> menu you can set the following:

Setting	IC-9700 default	ZL3DW setting	My Setting
Sidetone Level Turn it all the way down if you don't want sidetone.	50%	50%	
Sidetone Level Limit This apparently disables the sidetone if you turn the volume control up past the sidetone level.	ON	ON	
Keyer Repeat Time Sets the time before a memory message is re-sent if you have selected the automatic option. A red 'wait' icon is displayed between transmissions.	2 seconds	2 seconds	
Dot/Dash ratio The default is a 'Dah' that is three times the length of a 'Dot.' But this can be adjusted between 2.8 and 4.5.	**1:1:3.0**	**1:1:3.0**	
Rise Time Sets the CW rise time. Select 2 ms, 4 ms, or 8 ms. The default is 4 ms.	4 ms	4 ms	
Paddle Polarity can be set to normal or reverse to suit your operating style.	Normal	Normal	
Key Type can be set to Paddle, Bug, or Straight Key.	Paddle	Paddle	
MIC Up/Down Keyer allows the up-down buttons on the hand microphone to send CW. Up (left) is Dits. Down (right) is Dahs.	OFF	OFF	

DECODE

The DECODE icon is visible when the radio is in RTTY mode. It activates the internal RTTY decoder and eight message memory keys. You can also use four buttons connected via the microphone connector to send the first four messages. Set <MENU> <SET> <Connectors> <External Keypad> <RTTY>.

➢ DECODE sub-screen and Soft Keys

To edit an RTTY message, select RTTY then select <MENU> <DECODE> <TX MEM> <EDIT>.

- Touch the message that you want to edit and then touch the <EDIT> icon. A keyboard will appear on the touch screen.

- Messages can be up to 70 characters.

- Inserting a carriage return character ↵ causes the following text to appear on a new line at the receiving station.

- Press ENT to save your changes.

Soft Key	Function	Hold	Menu Setting / Notes
< 1 >	Change to Menu 2 selections.	None	
HOLD	Touch HOLD to halt the decoder. This stops text scrolling so you have time to read a callsign etc.	None	
CLR	None	Clear	Touch and hold to clear the received text area.
TX MEM	Opens the eight RTTY message memories.	None	They can be edited by touching the EDIT icon.
< 2 >	Change to Menu 1 selections.	None	
LOG	Opens a sub-menu. You can turn on the logging function and change it between a	None	The text log records all decoded text and the text that you send. See the LOG table below.

	text file and an HTML file.		
LOG VIEW	Opens a file dialogue where you can select an RTTY text log to view.	None	The text log records all decoded text and the text that you send. Useful data for your station log entries.
ADJ	Touching ADJ allows you to use the Main VFO knob to adjust the decode threshold. Touch ADJ again to exit.	None	The default threshold is 8. Even a setting of 15 does not stop occasional false decodes.
DEF	Touch and hold DEF to return to the default decode threshold of 8. Touch ADJ to exit.	Function	DEF is only visible on the ADJ sub-menu.
EXPD/SET	Increases the size of the Decoder screen so that the send and received text areas are larger. This overlays the Sub receiver display. But touching EXPD/SET again restores it.	SET	SET opens a menu with display and decoder options.

➢ The RTTY decode LOG submenu.

Setting	IC-9700 default	ZL3DW setting	My Setting
Decode Log When the Log is on, all decoded characters are recorded until you go back to this menu and stop it. A red dot in the top left of the RTTY Decode screen, next to the TX icon, indicates that the log is recording.	OFF	OFF	
File type	Text	HTML	

Text or HTML. HTML is better because it is in color.			
Time Stamp Adds a timestamp date and time at the start of each receive and transmit period. The date format is YYYYMMDD	ON	ON	
Time Stamp (Time) Sets the timestamp to local time or UTC. Note that this will be affected if you have set the radio clock to UTC. In my opinion, all logs should be in UTC, but it's up to you.	Local	UTC	
Time Stamp (Frequency) Adds the operating frequency to the log. "Yeah, why not?"	ON	ON	

➢ The RTTY decode SET submenu.

Touch and hold any menu item to reset the item to the Icom default.

Setting	IC-9700 default	ZL3DW setting	My Setting
FFT Scope Averaging Adding averaging makes the tuning spectrum scope smoother, but the delay that is introduced makes tuning in RTTY signals more difficult. Can average 2, 3, or 4 sweeps.	OFF	OFF	
FFT Waveform Color The default is pale blue, but you can adjust it if you want to. Some folks like a green or a greyscale image.	R:51, G:153, B255	R:51, G:153, B255	

Decode USOS	ON	ON	
USOS stands for unshift on space. Standard Baudot code will stay on characters or letters until a shift character is received. This function assumes that a letter is more likely to follow a space than a number. It automatically shifts the receiver back to letters after a space character has been received. Don't change this setting unless you are having decoding problems or are receiving secret messages sent in 5 number groups.			
Decode New Line Code	CR, LF, or CR+LF	CR, LF, or CR+LF	
This changes the decode screen. **CR+LF** sets a new line only when a carriage and line feed is received. The **CR, LF, or CR+LF** option sets a new line if a CR or a LF or both CR+LF character is received. The default option usually makes the received text easier to receive. But it can break up sentences.			

METER

Selecting <MENU> <METER> displays the Multi-Function meter. You can also enable the Multi-Function meter by touch and holding the meter display on the touch screen.

The Multi-Function meter works best with the sub-receiver and the spectrum scope turned off. That way you get the larger S meter/RF Power meter.

While receiving, the Multi-Function meter shows you the S meter, supply voltage, and temperature of the final MOS-FET stage. When the radio is transmitting the meter shows all the transmit metering at the same time. This is very handy when you are setting the Mic Gain and voice Compression controls.

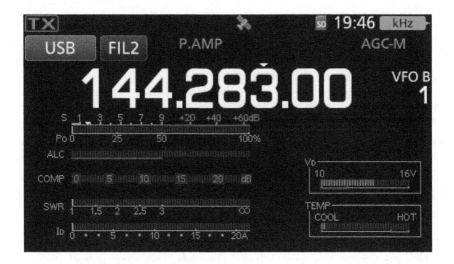

- S units (S0 to S9 to S9+60dB) [while the radio is receiving]
- Po (RF power output 0 – 100%)
- ALC (voice peaks should always be within the red zone)
- COMP (with the compressor turned on voice peaks should always be within between 5 and 20 dB. Never more than 20 dB)
- SWR (Standing wave ratio. Hopefully, less than 1.5)
- I_D (current being drawn by the final amplifier MOS-FETs)
- V_D (drain voltage on the final amplifier MOS-FETs)
- Temperature (measured at the final amplifier MOS-FETs).

<MENU> <SET> <Display> <Meter Peak Hold> adds a peak hold function to the S meter while receiving and the Po output power meter while transmitting. The default is ON.

SATELLITE

The Satellite Soft Key turns on the 'Satellite' tracking mode and the VFOs will change to the last used satellite frequency pair. The upper VFO becomes receive only. It is used for receiving the downlink from the satellite. The lower VFO, which must be on another band, is used for transmitting the uplink signal to the satellite. While not transmitting you can listen to the uplink frequency.

If you want to use the same band for receiving and transmitting during satellite operation, for example, to talk to Astronauts or Cosmonauts on the International Space Station, or when using a transverter, you can use the split function in the normal transceiver mode.

Three new Soft Keys are displayed. The top one (MAIN) links or unlinks the tracking of the downlink (receiving) VFO. The bottom one (SUB) links or unlinks the tracking of the uplink (transmitting) VFO. The middle one (NOR/REV) selects 'normal' or 'reverse' tracking when the VFOs are linked.

With 'reverse tracking' turned on, the receiver frequency will decrease as the transmit VFO is tuned higher and vice versa. You can select and control either the transmit or the receive VFO. This mode is used when operating through 'reverse tracking' SSB transponders. With 'normal' tracking enabled, the receiver frequency will increase as the transmit VFO is tuned higher and decrease as the transmit VFO is tuned lower. The normal-tracking mode is used when operating through FM, Digital, or 'normal tracking' SSB, transponders.

The satellite mode has its own set of 99 memory channels. They are accessed and stored in the same way as the normal 'band' channels. There are no 'call' channels or scan preselects in the satellite frequency bank.

Because satellite ground station operation is one of the most important functions of the radio, I have included a whole chapter dedicated to the satellite mode. Page 184.

MEMORY

The MEMORY Soft Key opens the main memory screen. You can save the current VFO frequency including the current mode and filter, delete a stored memory, or edit the name of an existing saved memory.

Rather strangely unless you happen to be in Memory mode rather than VFO mode, touching a memory returns you to the main operating screen but does not change the VFO to the selected memory frequency.

The radio can store 99 frequencies per band, including the mode and filter that was in use, plus three scan edge memories 1A-1B, 2A-2B and 3A-3C, and two 'Call' channels. There is also the facility to assign a scan group to any saved memory channel. The memory scan mode can scan the memory slots in a nominated scan group or all of the memory slots.

The satellite mode has its own set of 99 memory channels. But no 'call' channels or scan preselects.

There is also a memory bank for GPS data and the DV/DD memory bank for Digital Voice repeaters. This memory bank is covered in more detail in the D-Star chapter.

➤ MEMORY sub-screen Soft Keys

You can store the current VFO frequency to a memory slot by touch and holding any of the memory slots. If you select an already occupied slot you have the choice of overwriting the contents with the current VFO settings.

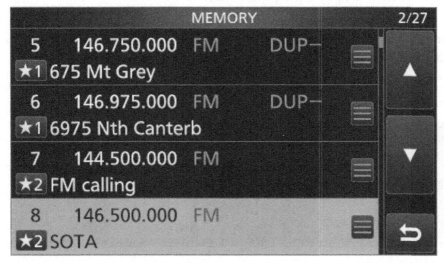

A standard memory bank

Each memory entry looks like the image above.

- You can use the up and down Soft Keys, turn the MULTI knob, or use the up ▲ or down ▼ buttons to display the 99 memories on pages 1 to 26.

- The number on the left is the memory slot number from 1-99. Or 1A-1B, 2A-2B and 3A-3C for the scan pre-sets, or C1 and C2 for the two 'Call' channels.

- The *1 or *2 is a scan group number. There are three scan groups. You can scan through any of the scan groups, all three scan groups, or all 99 memory slots. Touch the star to cycle through; *1, *2, *3, or none.

- Touch and hold a memory entry or touch the 'file' icon on the right, to select a memory slot.

- If the memory slot is blank there will only be one choice. You can save the current VFO settings into the memory slot, or not.

- If the memory slot is already programmed, you get three choices. You can add or edit the memory name. You can overwrite the current information with the current VFO settings, or you can clear the slot. You don't have to clear the slot before you overwrite it. Exit the edit menu using the ↻ Soft Key.

- When the slot has been programmed you can touch the star icon under the memory slot number to place the frequency into one of the three memory scan groups. Cycles through *1, *2, *3, and no scan group. Touch and hold clears the whole scan group or all three scan groups! Be careful.

➢ The alternative memory management system

You can also save memory contents using the memory submenu accessed by touching the memory number on the main touch screen. I am puzzled as to why the radio has two completely different methods of managing the memories.

For more on the alternative memory management system see 'Saving a frequency to a memory slot' on page 94 and 'Recalling a stored memory channel' on page 95.

The quick way to save a frequency into a memory slot

➢ FM Repeater Channels

There is no specific memory storage for FM channels. But you can use the standard memory channels and name the memory channel with the repeater identification. The standard memory channels will store the repeater duplex offset and CTCSS tone so set them before saving the channel to a memory slot.

➢ External memory managers

The IC-9700 Programming Software by RT Systems and the free CS-9700 memory manager from Icom can read memories and menu settings from the radio. You can edit the memory banks in the PC software and then upload them back to the radio over the USB cable. The program uses the 'clone' mode over the USB interface. Both programs allow you to set most (maybe all) of the menu settings and you can store a copy of the radio configuration on your PC.

Because I make quite a few changes, I always download a fresh copy of the radio configuration from the radio, make any changes that I want using the software, store the resulting file to the PC, and then upload the revised configuration back to the radio.

SCAN

Selecting <MENU> <SCAN> displays the SCAN Menu. Operating and configuring the Scan mode has been covered in 'Operating using the scan mode' on page 90.

The front panel SCAN button has slightly different functions including some neat memory slot scanning options. See SCAN button on page 64.

The SCAN mode also works when you are in the DR D-Star digital repeater mode. It scans the DD/DV memory bank for repeaters within the range of the radio. See D-Star (DR) Scan Mode on page 176.

MPAD

Selecting <MENU> <MPAD> displays the MEMO PAD menu. The memory pad is used for short-term storage of a frequency and mode that you might want to return to later. As new entries are added to the top of the stack, the oldest disappear. The memory pad can store either 5 or 10 frequencies, depending on the setting, <MENU> <SET> <Function> <Memo Pad Quantity>.

I am not sure why Icom bothers to include this sub-screen. It gives you an opportunity to see what is saved in each memory pad slot, but the next time you press and hold the MPAD button to store a new temporary record, all the entries will move down one slot. Pressing the MPAD button repeatedly cycles through each stored frequency. The screen does give you the option of deleting one or all the stored memories... but why bother?

Soft Key	Function	Hold	Menu Setting / Notes
▲	Step up through the stack and a "temporary Pad."	None	The MPAD stack holds 5 (or 10) channels. Up/Down cycles through them plus the "temporary Pad."
▼	Step down through the stack and a "temporary Pad."	None	
DEL	none	Delete entry	Touch and hold to delete the currently highlighted entry.
DEL ALL	none	Delete all entries	Touch and hold to delete all of the memory pad entries.

EXPAND	Increases the size of the MEMO PAD so that you can see 6 entries at once.	None	

RECORD

The QSO recorder records the receiver and transmitted audio. You can start recording by selecting <<REC Start>> on this screen, or via the QUICK button.

A pause **II** symbol above the VFO/MEMO icon indicates that the recorder is paused waiting for the receiver squelch to open. A red dot ● symbol indicates that the recorder is recording. Also, the blue SD card icon will flash indicating a 'write' to the SD card. A recording will start automatically when you transmit by pressing the microphone PTT switch.

To stop recording press the <QUICK> button and select <<REC Stop>>, or you can go back into the <MENU> <RECORD> and select <<REC Stop>>.

Soft Key	Function	Hold	Menu Setting / Notes
<<REC Start>>	Touch <<REC Start>> to start recording audio to a file on the SD card.	None	● REC means recording II REC means paused (squelch closed)
<<REC Stop>>	Touch <<REC Stop>> to stop recording	None	
Play Files	Allows you to select a recorded file and play it back.	None	No, you cannot use the recording to modulate the radio. Touch and hold a folder or a file to delete it.
Recorder Set	See the next table.	None	
Player Set	Sets how far the audio will skip ahead if you touch the fast forward icon during playback. You can skip back too.	None	3, 5, 10, or 30 seconds. Touch and hold sets the time back to default 10 seconds.

> Recorder Set – sub-menu

Setting	IC-9700 default	ZL3DW setting	My Setting
TX REC Audio. Record microphone audio (Direct) or the Transmit Monitor audio (Monitor).	Direct	Direct	
RX REC Condition. Record only when the receiver squelch is open or record all the time.	Squelch Auto	Squelch Auto	
File Split. Create a new file when you transmit, or the squelch opens (ON). Or keep recording until manually stopped or the 2Gb maximum file size is reached (OFF).	ON	ON	
REC Operation. MAIN/SUB linked or separate. If set to 'MAIN/SUB link' the recorder will record both receivers if either is un-squelched. If 'Separate' each channel records if un-squelched.	LINK	SEPARATE	
PTT Auto REC. If set to 'ON' a recording will start every time you transmit. If 'OFF,' recording will only start if you start it.	OFF	OFF	
PRE-REC for PTT Auto REC. If 'automatic recording' is ON, Pre-Rec will record up to 15 seconds of the received audio before you press the PTT on the microphone.	10 seconds	10 seconds	

The second page of MENU icons relate to the radio's digital voice operating modes.

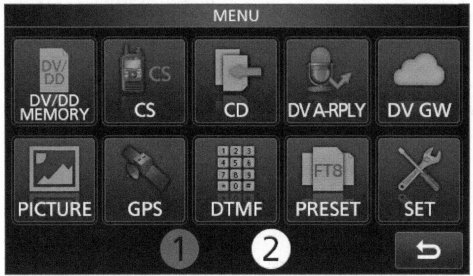

The main MENU (page 2 of 2).

DV/DD MEMORY

The DV/DD memory 'Repeater List' is used to store D-Star digital voice and data channels for all three bands. It can hold 2500 records divided into 50 repeater groups. This seems like a lot, but the radio can access D-Star and other digital voice repeaters worldwide through local repeaters or hotspots which are linked to worldwide networks via the Internet. Obviously adding all of these channels by hand would be a major hurdle, so the radio has the capacity to load the bank from a CSV file stored on the SD card. You can also use the CS-9700 or RT Systems programmers.

The Icom manual states that the DV/DD memory bank is pre-programmed, but my radio was supplied blank. You can download a file containing the D-Star repeaters from the download area of the Icom Japan website.

http://www.icom.co.jp/world/support/download/firm/IC-9700/repeater_list/.

I modified the data file to put the New Zealand repeaters into Group 01 so that they appear at the top when I want to use them.

I also put my local D-Star repeaters at the top of the Group 01 list. You can include normal FM repeaters in the list if you want to.

Copy the file onto the IC-9700\CSV\RptList directory of the SD card and then import the file using <MENU> <SET> <SD Card> <IMPORT/EXPORT> <IMPORT> or use the IC-9700 or RT Systems programmer.

The DV/DD memory sub-menu is also where you set the name and call sign of your buddies that you will call a lot using the D-Star mode. <MENU> <DV/DD memory> <Your Call Sign>. Yes, the title "Your Call Sign" is a bit misleading. Perhaps it means 'your list of commonly used call signs?' Touch and hold the highest available entry line to add or edit a call sign and name.

If you touch and hold a memory entry, you can add, edit, move, or delete DV/DD memories. Or skip them during the DR scan. The 'Skip all ON' setting skips the entire group. 'Skip all OFF' includes the entire group in the DR scan.

If you touch and hold a memory group title, you can edit its name or alter the skip setting to skip the entire group, or to include the entire group. There is no point scanning for repeaters in different countries or out of the range of your radio. The only channels that you want to be included in the DR scan are your local repeaters. I found the easiest thing to do was to set 'Skip all ON' on all of the groups and then use the individual SKIP Soft Key to set my four local repeaters so they will not be skipped. In DR mode, press the SCAN button to scan the FROM repeater through your local channels. They can include FM repeaters as well as D-Star repeaters.

If you hear a signal from a linked repeater or gateway that is not in the memory list. You can add the repeater to the list of repeaters in the DD/DV memory from the RX History screen. See the CD sub-menu below.

CS

The CS menu shows you the data field settings for the D-Star DR mode. The Field Structure is covered in detail in the D-Star chapter, page 165.

➢ UR, 'Your Call' holds the information being sent to the repeater and the Internet. It is the information in the TO box. Normally it will be set to 'CQCQCQ,' but it can be I for information, U for unlink repeater, or E for Echo. It can also contain the call sign of a station you want to work, a gateway repeater, or a reflector.

   Touch and hold can be used to add a repeater listed in the TO box to the DD/DV repeater list.

➢ R1, 'Repeater 1' is set to the call sign of the repeater that you are calling through. The call sign is followed by a letter which must be the eighth character. For example 'ZL3DV B' would be a 70 cm repeater in New Zealand.

   It is the information in the FROM text box.

Module	Function
A	23 cm repeater
B	70 cm repeater
C	2 m repeater
D	Digital Data

If R1 is blank or (NOT USED*) it means that you are making a Simplex Call. No local repeater is selected.

Touch and hold can be used to add a repeater listed in the FROM box to the DD/DV repeater list.

TECHNICAL TIP: Japan uses A for 70 cm repeaters and B for 23 cm repeaters. Some systems have multiple modules on the same band, so B, C, and D might all be 70 cm repeaters. It does not really matter so long as you have the correct letter code for the repeater you will be accessing with your radio. If the repeater call sign is correct but the letter is incorrect you will not get a message back when you transmit. In that case, try a different letter.

➤ R2, 'Repeater 2' is set to the call sign of the gateway. The call sign is always followed by the letter G, for gateway. The G must be the eighth character. For example 'ZL3DV G' would be the gateway for the ZL3DV repeater.

Module	Function
G	Gateway

If R2 is blank or (NOT USED*) it means that you are making a Local Repeater Call. No gateway repeater is selected.

There is no touch and hold function.

➤ MY, 'My Call' holds your call sign including the /9700 suffix that you set in D-Star First Steps. The field is eight characters before the / symbol and four characters after the / symbol.

Touch and hold to change from one version of your callsign to another, or edit the data that will be sent.

CD

The CD menu opens the D-Star RX HISTORY sub-menu. You can see information about the stations that have been received in the D-Star mode. The information includes the callsign and radio identifier, any received information text the repeater, the date and time of the call, whether the call was via a gateway or direct, what was in the 'TO' box at the time, and the GPS range and direction if a GPS packet was included in the transmission.

The DETAIL Soft Key adds five additional pages of information about the received caller. If a GPS packet was included in the transmission this can include the location latitude and longitude, Maidenhead grid, altitude, distance, course, and speed of the received station.

The RX>CS Soft Key loads the appropriate repeater (gateway) information into the R1 and R2 data fields and the callsign of the station into the TO box (UR data field), so that you can call the station.

DV A-RPLY

The DV auto-reply is a voice message for the D-Star DR mode. Recording the message is similar to recording one of the voice-keyer messages.

When a D-Star call specifically addressed to your call sign is received, the radio will normally send back your call sign data string. If you have the 'Voice' option selected it will also send the 'auto-reply' voice message.

- To turn the voice auto reply option on, select <MENU> <SET> <DV/DD SET> <Auto Reply> to Voice.

- To turn the text auto reply option on, select <MENU> <SET> <DV/DD SET> <Auto Reply> to ON.

- To turn the text auto reply option off, select <MENU> <SET> <DV/DD SET> <Auto Reply> to OFF.

DV GW

The Digital Voice Gateway option enables an Internet connected radio to act as a D-Star gateway or HotSpot. I have no experience with setting this function up, but there is a big section about it in section 11 of the Icom Advanced manual.

➢ The IC-9700 Terminal Mode

The terminal mode lets you configure the radio to access gateway repeaters and reflectors directly over the Internet, without using a repeater or any radio transmission at all. This is not really ham radio in my opinion. More like Skype. You need an Ethernet connection and you may have to open a port on your Internet router to let the radio access the Internet.

> The IC-9700 Access Point Mode

The Access Point mode is like the terminal mode except that it acts as a hot-spot. The radio is connected via the Ethernet cable and the Internet to a gateway or reflector. You can use a handheld radio to talk to the receiver in the IC-9700 which passes the audio through the Internet to the distant gateway. Audio from the gateway is transmitted by the IC-9700 to your handheld radio. This is a very expensive way to implement a D-Star hot-spot, but I guess it could be useful if you wanted to make D-Star calls from around your yard.

> OPC-2350LU

If you can't connect the radio directly to an Ethernet port on a fiber or ADSL router connected to the internet, you can use an OPC-2350LU or similar data cable to connect the radio to a Windows PC or Android device and then use that arrangement to connect the radio to the Internet. A Windows device will need RS-MS3W software and an Android device will need RS-MS3A software. The software is a free download.

PICTURE

The picture sharing mode was introduced in firmware update 1.20. It allows you to send and receive photos over the D-Star simplex or repeater mode. You can take a photo with your phone, copy it to your IC-9700 via Wi-Fi and the radio's Ethernet connection, and transmit it to another station. "Here is a photo of us at the beach!" or "This is my latest toy, an icom IC-9700!" Picture sharing is fully described in the D-Star section on page 177.

GPS

GPS location information is used by the radio for three things. It can be transmitted with your D-Star transmissions so that the station you are calling can see; your location, Maidenhead grid reference, distance, speed, the direction you are heading, and what direction you are from them. It is also used for the 'Near Repeater' function which lets you choose DV or FM repeaters that are near your location. And it is used to determine the distance and heading towards stations that you hear via the repeater or gateway.

If you want to run D-Star but you don't plan to operate mobile or portable, there is little point in connecting an externals GPS. Just enter your location manually.

➢ GPS Set

GPS Set is used to select; OFF if you are not using D-Star at all, 'Manual' if you don't have a GPS receiver attached, or 'External GPS' if you do have a GPS receiver attached.

➢ GPS TX Mode

GPS TX Mode set whether you will transmit your position with your D-Star transmissions. And if so, whether the data is sent in NMEA or D-PRS format. A sub-menu determines which information options are included in the transmission, (see the Icom Advanced Manual 9-19). You can also add a 20-character text string to go with the location information.

There is more information about the GPS functions in the GPS chapter, page 180.

➢ GPS Information

The GPS Information screen shows the satellite number and bearing of satellites being received by the connected GPS receiver. The image is a 'Radar Plot.' The center is your location. Satellites that are high overhead are shown close to the center. The outer ring is the horizon. The white dots are GPS satellites that are being received but the data is invalid, usually because of extreme range. If no GPS receiver is connected, there is no display.

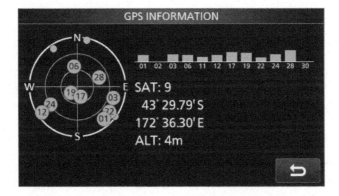

This screen can also be accessed via the QUICK button or by touching the GPS icon on the touchscreen..

➢ GPS Position

The GPS Position screen shows your location. Latitude, longitude, Maidenhead Grid, altitude (can be inaccurate), your speed, course, and the current time.

The second screen shows the GPS data from a received D-Star signal.

The third screen can show the GPS data from a location stored in the GPS memory bank. Touch and hold the screen to load a location from the GPS memory.

The fourth screen can show the GPS Alarm distance and bearing to a nominated location. When the alarm is set, a beep is issued by the radio when you drive within 1 km of the nominated site or group of GPS locations. At 500 meters three beeps are issued. This is a function known as 'geo-fencing,' (advanced manual page 9-26). It can also be used to alert you when a 'target' station in your list comes within 1 km or 500 meters of your location or within the selected area. The station has to transmit within the area for the alarm to work.

This screen can also be accessed via the QUICK button.

➢   GPS Memory

The GPS Memory screen provides access to the GPS memory bank. The bank can store 300 GPS locations divided into 26 groups.

You can add positions manually, but it is easier to use the CS-9700 memory manager or load them from a .csv file on the SD card. You can download a list of GPS locations in .csv format from the Icom Japan downloads area. It is bundled with the D-Star repeater list.

http://www.icom.co.jp/world/support/download/firm/IC-9700/repeater_list/

You can edit a .csv file using Excel or another compatible spreadsheet program.

Place the SD card into your PC card reader. Copy the file onto the IC-9700\CSV\GpsMemory folder on the SD card and import it using <MENU> <SD Card> <Import/Export> <Import> <GPS Memory>.

➢   GPS Alarm

GPS Alarm is used to set the GPS alarm geofencing parameters. This is quite complicated to set up and I believe that few owners will bother using the capability. The navigation features of your cell phone or in-car navigation system are far superior. See Icom Advanced Manual 8-26.

The GPS alarm can also be used to alert you when a 'target' station in your list comes within 1 km or 500 meters of your location or within the selected area. The station has to transmit within the area for the alarm to work. It could be in a QSO or using the GPS Auto TX beacon mode.

The GPS Alarm function is indicated with a blue icon on the top row of the display, beside the GPS icon if an external GPS receiver is connected.

➢ GPS Auto TX

GPS Auto TX can be set to send your location data over D-Star at regular intervals ranging from every 5 seconds, which is far too often, to every 30 minutes. Every three to five minutes should be OK for a mobile station. Any more than that and you may annoy other repeater users. It is rather like sending APRS beacons on an FM repeater. The format will be either NMEA or D-PRS (D-Star packet reporting system), depending on the <GPS> <GPS TX Mode> setting.

DTMF

Many D-Star and some FM repeaters can use short DTMF codes to control repeater functions including linking to reflectors. You can send a DTMF code by using the DTMF screen. <MENU> <2> <DTMF>. Touch SEND to immediately send a code using 'Direct Input,' or you can send one of the last codes sent. Touch EDIT to pre-load up to 16 saved codes. SET changes the speed that the codes are sent from the transmitter. I have changed to 200ms because my repeater seemed to miss the codes when the speed was set to 100ms. But you can try 100ms first.

There is more information about DTMF code strings functions in the D-Star chapter, page 175.

PRESET

The February 2021 (V1.30) firmware update added a "one-touch FT8 Preset mode" to the second page of the main menu. It adds five one-touch "mode pre-sets." The idea was to quickly change the radio to the settings required for FT8, reflecting the popularity of the mode. The top pre-set is labelled 'Normal,' but I have renamed it to 'SSB' because it does not return the radio to the previous non-Preset setting. The second item is labelled 'FT8,' although the options that it sets are suitable for most external digital modes. You can edit those two settings and add three more pre-set arrangements of your own design.

Load the FT8 Preset before you start WSJT-X or another digital mode program. <MENU> <2> <PRESET> <FT8> <YES>

Unload it again when you have finished with the FT8 or other Preset.

<MENU> <2> <PRESET> <UNLOAD> <YES>

The full description of the FT8 Preset mode starts on page 32.

SET MENUS

<MENU> <SET> is used to access all the deep menu settings. I have covered the important ones that you will probably need to adjust in the 'SPECIAL SET MENU ITEMS' section below.

These include

- The USB 'send' and keying settings (RTS and DTR) control lines for PTT and CW

- The CI-V settings (data rate and Icom address). There is no COM port allocation as they are set by the Icom driver software, and

- Audio level settings (for the USB cable)

The following is a summary of the other SET Menu adjustments. Many of them are discussed more fully in the 'Setting up the radio chapter' and throughout the book.

➢ Tone Control/TBW

- RX. Set audio high pass and low pass filters for the transmit modes. Or set the bass and treble for the AM, FM, and SSB modes. See page 19

- TX. Set the bass and treble for AM, FM, and SSB. Set transmit bandwidth for SSB and SSB Data only. See page 18

➢ Function

- Beep settings (six different settings)

- User Band Edge. Set three band edges for each of the three bands. This is a hidden function, see page 47.

- Sub Band Mute TX. Selects whether you want to mute the sub-receiver while you are transmitting. You can choose to mute the speaker, USB audio out, and/or the LAN audio output. Default is not muted.

- RF/SQL Control. (RF and Squelch, Squelch only, Auto). Page 53

- FM/DV Center Error. (Enables flashing the S meter green LED for off-frequency signals)

- TX Delay. Transmit delay from PTT to transmit. Set to at least 10ms if you are using a linear amplifier. Longer if it is relay switched. Otherwise OFF.

- Time-Out timer. Stops the rig transmitting forever if the microphone gets wedged down the back of the seat of your vehicle. You should definitely set this if you operate mobile.

- PTT Lock. Designed to stop you from transmitting accidentally. 'ON' stops the transceiver from transmitting.

- Split settings. Quick SPLIT, SPLIT Offset, SPLIT LOCK. See page 84

- Auto Repeater. May not work in your region. See page 40.

- RTTY settings

- SPEECH button settings

- Dial LOCK button settings

- Memo Pad Quantity, 5 or 10 memory slots

- Main Dial Auto TS. Increases tuning step if you move the VFO knob quickly

- Mic Up/Down scan speed

- AFC Limit. Limits the effect of the AFC according to the FM filter bandwidth. Default ON.

- Notch Filter auto/manual SSB mode

- Notch Filter auto/manual AM mode

- SSB/CW synchronous tuning. Sets the VFO to the CW offset when receiving CW on SSB mode. The default is OFF

- CW 'normal side' LSB or USB (default USB)

- Onscreen keyboard type (QWERTY or 10 key phone pad)

- Full keyboard layout (English, German, or French). This may be different on your radio due to market zones

- Screen capture settings

- Reference frequency adjustment. Locks the main oscillator to an external 10 MHz reference oscillator. You probably should not touch this unless you have a precision frequency reference connected. But you can adjust the settings manually to net the radio to a known input signal such as a beacon.

➤ My Station

The 'My Station' menu is used to set the basic parameters for D-Star operation. Your call sign, a message to go with your transmission (usually your name and city or district), and your call sign for the digital data mode. This usually defaults to the same call sign as used for the DV mode.

➤ DV/DD Set

The DV/DD Set menu sets the data transmission parameters for D-Star operation. I have left them all set to their default settings.

- Standby Beep. ON makes a beep when a DV signal disappears. 'On to me' makes a high pitch beep when the radio receives a signal that is addressed to your call sign. 'On to me: Alarm' sounds an alarm when a received signal carrying a transmission addressed to your call sign disappears.

- Auto Reply turns the automatic reply function to ON, OFF or Voice

- DV Data TX. If set to AUTO sends the data from a PC over the DD channel. If set to PTT it waits until you press the PTT before sending the data signal. There are other menu settings required including GPS select and USB B/DATA. See the Icom Advanced manual 10-22.

- DV fast data. Fast data over the DV D-Star voice mode can be received by some models of Icom radio running the latest firmware. Fast data is 3480 bps. Normal data is 950 bps.

- Digital Monitor (selects the receiver mode when XFC is pushed while in DV mode. The options are Auto (DV or FM), Digital (DV), or Analog (FM).

- Digital Repeater Set automatically sets the transceiver FROM box to match an incoming signal from a repeater. Default ON.

- DV Auto Detect automatically switches the radio from DV to FM when an FM signal is received. For example, while scanning.

- RX Record (RPT). The radio can record data from the 50 most recent DV mode received transmissions (ALL), or just record the most recent data (Latest Only).

- BK allows you to break into a conversation where two stations are using digital squelch.

- EMR (enhanced monitor request) is used only in an emergency situation to call all stations on the repeater. The audio voice signal is heard at the volume set by EMR AF Level, even if the far station has the volume turned down.

- EMR AF Level. Sets the level of the emergency voice message transmission.

- DD TX Inhibit (Power ON) sets the ability to transmit while in the DD mode. It can also be turned on or off by pressing the TRANSMIT button. See the DD Mode in the Icom Basic manual section 12-1.

- DD Packet Output. Normal (default) outputs data packets when a call addressed to your call sign a CQCQCQ call is received. Or when emergency 'EMR' signal is received, or someone breaks in with a 'BK' signal. 'ALL' sends all packets.

➢ QSL/RX Log sets the logging functions

- QSL Log ON or OFF sets whether the radio will store a log on the SD card. The log is stored in the IC-9700\QsoLog directory in Excel .csv format. It stores a log of all transmit periods in any mode, recording the 'time on' on one line and the 'time off' on the following line. It will work in any mode, but the full data set is only captured if you are using the DV mode.

- RX History Log ON or OFF sets whether the radio will store a log of received signals on the SD card. The log is stored in the IC-9700\RxLog directory in Excel .csv format. This mode only works in the DV D-Star mode.

➢ Connectors

- External P.AMP selects whether you are using an external preamplifier. There is a setting for each band. **WARNING** turning this setting on will apply a DC voltage to the antenna connector for the band you have selected. The voltage is used to power the preamplifier over your coaxial feeder coax. **Make sure that your antenna does not present a zero or very low impedance to DC before applying the volts. In other words, don't turn this on unless you are using an external power amplifier that is powered over the coax cable.**

- External Speaker Separate. 'Separate' means that the 'Main' receiver output is sent to the EXT-SP MAIN speaker jack on the rear panel of the radio. The 'Sub' receiver output is sent to the EXT-SP SUB speaker jack. 'Mix' outputs a mix of both receivers to both speaker jacks, (the same as the internal speaker).

- Phones: is used to set the audio level to your headphones. The level is controlled by the volume controls, but this setting allows you to make the audio level the same in the phones as it is in the speaker. (0) is nominally the same level as the speaker. You can also select what is sent to each earpiece. 'Separate' puts the 'Main' receiver output on one side and the 'Sub' receiver output on the other. 'Mix' puts a mix of the 'Main' and b' receiver into both sides (mono) like the internal speaker. Auto selects separate when both receivers are enabled and the 'Main' receiver on both sides when only the 'Main' receiver is enabled.

- ACC AF/IF Output sets level and squelch settings for audio to be sent out of the ACC jack.
- USB AF/IF Output sets level and squelch settings for audio to be sent out of the USB cable.
    - Output Select AF (audio) or IF (12 kHz) output
    - AF output level
    - AF squelch. The default is 'OFF' so the audio to the computer is always available. The 'ON' mode only sends audio to the computer if the squelch is open
    - AF Beep/Speech. OFF: beeps and SPEECH button audio are not sent to the computer. ON: beeps and SPEECH button audio are sent to the computer
    - IF output level
- MOD INPUT
    - ACC MOD level. Audio input from peripheral on ACC connector
    - USB MOD level. Audio input from PC on the USB cable
    - LAN AF/IF Audio input from PC on the LAN cable
    - DATA OFF MOD. Sets the source of modulation audio when the radio is not in a data mode. The default is MIC, ACC. If you set USB as one of the settings, you can transmit audio from the PC in voice modes. This is a mixed blessing. It lets you transmit message macros from N1MM Logger etc. But the microphone will be live, and the compressor, VOX, and audio bandwidth settings are in use, which may not be optimal for digital modes
    - DATA MOD. Sets the source of modulation audio when the radio is in a data mode. The default is the USB cable
- ACC SEND Output. Allows you to select which bands output a SEND (PTT) low signal on the ACC jack when you transmit. For example, you may have a linear amplifier that works on the 2m band but not on the other two bands. You can set the 2m band to ON and the other two bands to OFF
- USB Send Keying sets the COM port RTS and DTR settings. See page 30.
- External keypad enables the sending of message memories using buttons connected to the microphone connector, (Voice, Keyer, RTTY)
- CI-V. See the CI-V settings table on page 138
- USB (B)/DATA function.

o   USB (B) Function sets what the second COM port (USB port B) will be used for. Nothing, RTTY decoded output, or DV low-speed data input and output.

o   DATA Function sets what the 2.5mm DATA jack will be used for. A GPS Receiver, RTTY decoded output, DV low-speed data input, and output, or CI-V control.

o   GPS Out. Select whether you want to send the GPS data from a connected GPS receiver out of the USB (B) port. 'DATA Function' must be set to GPS/Weather and 'USB (B) Function' has to be set to DV data or OFF.

o   DV Data/GPS Out baud rate sets the baud rate of DV or GPS data

o   RTTY decode baud rate sets the baud rate of decoded RTTY data

➢   Network

You can set a static IP address, subnet mask, and gateway. But it is easier to just select DHCP and let the network allocate an IP address. As I don't use the RS-BA1 remote control, or static addressing (YUCK!), I left everything at default settings.

If you are using the RS-BA1 remote control software you can give the radio a network name, disable or enable remote control, enable or disable a shut-down or standby mode, set the UDB port for control data, set the UDB port for audio data, set up to two user names, and change the radio name to the network. See the Icom Basic manual 8-15 for details.

➢   Display

*   LCD Backlight sets the display brightness

*   Display Type sets a black (default) background or a blue background

*   Display Font sets either 'basic' (default) or 'round' an Arial type font

*   Meter Peak Hold: What it says on the box, (ON)

*   Memory name: Display or do not display the memory name, "there is no try." If both receivers are tuned on the memory name is only displayed when the panadapter display is turned off. (ON)

*   MN-Q Popup. Very odd! It sets whether the radio will show the manual notch width popup when you select manual notch

*   BW Popup (PBT). Also very odd! It sets whether the radio will show the PBT popup when you change the PBT controls

- BW Popup (FIL) Yep this one is odd as well. It sets whether the radio will show Filter popup when you change the touch the filter Soft Key

- Rx Call Sign Display. In the DV mode, this sets whether to display the call sign of a received station. You can set it to scroll once or stay on the display until the received signal disappears or scroll once and scroll again when the received signal disappears. This is like having a control to set how long the roof light stays on after you lock your car.

- RX Position Indicator. You can set whether you want the incoming GPS data to display the caller's position indicator.

- RX Position Display. You can set whether you want the incoming GPS data to display the caller's position data in a dialogue box. OFF, Main Receiver Only, or both Main and Sub receivers, (if both are in DV DR repeater mode).

- RX Position Display Timer. How long do you want the incoming position data to remain visible?

- Reply Position Display, selects whether the radio should display the incoming station's position data while sending an Auto-Reply message or voice message.

- TX Call Sign Display. Select whether to scroll display the call sign of the station you are calling, or your own call sign while transmitting. Note the "Your Call Sign" is the call sign of the station that you are calling. "My Call Sign" is your call sign. Hmmm OK.

- Scroll Speed. Sets the scroll speed of information that scrolls on the display. Slow is very slow.

- Screen Saver. Sets the timeout period for the screen saver. The default is 60 minutes. I set my radio to 30 minutes.

- Icom opening message and splash screen settings. Opening Message and Power ON check. See 'Display your callsign when the radio starts' on page 15

- Display Unit sets latitude and longitude format, metric or imperial measurements, and the data format for NMEA weather station data.

- System Language: English or Japanese (on my radio).

➢ Time Set

- Date and time setting, immediate NTP TIME SYNC (needs a LAN Internet connection), NTP Function (needs a LAN Internet connection), NTP Server Address, I am using time.nist.gov.

- The NTP Function uses the Internet to accurately set the clock. It operates a little while after the radio boots and at periods after that as well. See page **Error! Bookmark not defined.**.

- GPS Time Correct. If you have a GPS receiver connected for DV mode position data. This setting allows you to use the GPS data to set the clock rather than using a LAN connection just for access to the NTP server.

- UTC offset. Set the UTC offset for your location. Make sure that you account for daylight saving if it is currently in use.

> SD Card

  - Load Setting - reset the radio configuration from a stored backup

  - Save Setting – store the radio configuration as a backup. Select <<New File>>. You can change the file name, but the default name includes the date and a version number, so I would just touch the <ENT> Soft Key and then <YES>.

  - Save Form. This lets you save the configuration data in the current firmware format. But if you are planning on rolling back to an earlier firmware version, you have the option of saving the data in the old (now obsolete) data format. That will let you load the configuration file after loading the old firmware.

  - Import/Export. You can import or export the;

    o 'Your Call Sign' list (remember this is the list of your buddies that you call often in D-Star mode).

    o DV/DD Repeater List

    o GPS memory List

    You can adjust the .csv data format, but the default setting matches the data format used for files on the Icom Japan website.

  - SD Card Info shows the size and spare capacity of the SD card. It also shows the amount of recording time left on the card.

  - Screen Capture View displays stored screen capture images on the display. The Power LED flashes to show that the radio is in screen saver mode. Touching the screen closes the image.

  - Firmware update from the SD card. See page 44

  - Format the SD card. Icom recommends formatting the SD card or USB stick, using the function in the radio, before using it for storage. Page 57. I did format the card before using it for the first time with the IC-7610. Since then

I have used the card for the IC-7300 and the IC-9700 without re-formatting it. There is plenty of space on the card for data from all three radios.

- Unmount the SD card. This option is used to 'unmount' the SD card or USB stick before you remove it. The same as you would when removing a USB stick from your PC. Page 57. I don't bother unmounting the card and have not experienced any problems, but this is at your risk. **Do not unmount the SD card while the radio is recording or otherwise writing to it, i,e the SD indicator is flashing.**

➤ Etc. Others

- Information: Shows the firmware versions and the MAC address. I wonder why the MAC address is not in the Network section. The MAC address is unique to your radio. You could take a note of it along with the serial number in case the radio ever gets stolen. Unlike a serial number, the MAC address can't be changed.

- Clone puts the radio into clone mode. Since memory managers like the Icom CS-9700 and the RT Systems programmer do this automagically, I don't know why you would want to use this setting. Power down and restart the radio to get out of clone mode.

- Touch screen calibration. The Icom Basic Manual says to refer to the Icom Advanced Manual and the advanced manual does not include this setting. You should only play with this if you are having issues with the touch screen display such as part of it not working or not sensing touches in the correct places. Select < Touch screen calibration>.

  A small white dot will be displayed in the top left of the display. Touch each white dot as it appears. That's it! Once touch four dost and hear the double beep the calibration is complete.

- Radio Reset

  o 'Partial Reset' sets operating settings back to factory default. It does not erase memory channels, call sign data, keyer memories, band edges, or ref adjust. Try a partial reset first.

  o 'All Reset' sets everything back to factory defaults

TIP: If the radio freezes and none of the front panel buttons will work, you will not be able to access the <MENU> <Others> <Reset> menu.

You can try the following:

1. *First, I guess you should check that it is not locked. Press and hold the SPEECH/Lock button. If you get a popup saying, 'Dial Lock On' or 'Panel Lock On," Press and hold the SPEECH/Lock button again and you should get a popup saying, 'Dial Lock OFF' or 'Panel Lock OFF."*

2. *The next thing I suggest is power down the DC supply completely or unplug the DC cable. Don't just turn the rig off. Then power up again. That should give you a full boot. Do you see the usual splash screen and callsign when you boot the rig?*

3. *You can try a Full Hardware Reset. Turn off the transceiver. Hold down the **PBT** and **V/M** buttons and press POWER to turn the rig on. This "should" reset the rig. You will lose saved memories etc. but at least the rig will work.*

SPECIAL SET MENU ITEMS

This section covers the items buried in the menu structure that you will probably need to adjust. These include

- The USB 'send' and keying settings (RTS and DTR) control lines for PTT and CW

- The CI-V settings (data rate and Icom address). There is no COM port allocation as these are set by the Icom driver software

- Audio level settings (for the USB cable)

➢ USB SEND/Keying settings

Select <MENU> <SET> <Connectors> <USB SEND/Keying> to display the settings.

The USB 'send and keying' settings configure the way that PC digital mode software sends CW and PTT signals to the radio. You can use either line for the transmit PTT as long as you use the other line for CW.

I use RTS (ready to send) for the SEND [PTT] control and DTR (device terminal ready) for the CW. It is important that your digital mode software uses the same control lines. The USB Keying (RTTY) setting is for FSK RTTY operation (as opposed to AFSK RTTY). It should be set the same as CW. On my radio that is DTR.

For more information, see 'Radio and COM Port device setting' on page 26 and USB Send/keying settings' on page 30.

Setting	IC-9700 default	ZL3DW setting	My Setting
USB SEND	Off	USB (A) RTS	
USB Keying (CW)	Off	USB (A) DTR	
USB Keying (RTTY)	Off	USB (A) DTR	
Inhibit timer at USB connection	ON	ON	

The 'Inhibit timer at USB connection' function is used to stop the radio sending a SEND, transmit keying signal when the USB cable is first plugged in. The manual states that this is only a problem if you are using old firmware. But I left it set to the default ON setting.

➢ CI-V Settings

<MENU> <SET> <Connectors> <CI-V>

These settings are for the CI-V (CAT) connection between the radio and digital mode software running on your PC. I experimented and found that the default settings are best.

	Setting	IC-9700 default	ZL3DW setting	My Setting
1	CI-V Baud Rate	Auto	Auto	
2	CI-V Address	A2h	A2h	
3	CI-V Transceive	ON	ON	
4	USB/LAN→REMOTE	00h	00h	
5	CI-V USB Port	Unlink from REMOTE	Unlink from REMOTE	
6	CI-V USB Baud Rate	Auto	Auto	
7	CI-V USB Echo Back	OFF	OFF	
8	CI-V DATA Baud Rate	19200	19200	
9	CI-V DATA Echo Back	OFF	OFF	

1. CI-V Baud rate sets the data rate to a device connected to the REMOTE jack. If the 'Link to [REMOTE]' option is chosen, the CI-V Baud rate also sets the data rate to a PC connected via the USB port.

2. CI-V Address is the Icom address for the radio. The default is A2h. it should not be changed unless that is the only way that you can connect to a software package. Some earlier Icom radios used 88h and most radios don't use an address at all. The IC-7610 uses an address of 98h. You can change the address in the radio using the CI-V settings, but this would be a last resort because it would probably cause communication with other software that is expecting to use A2h to fail.

3. CI-V Transceive sets whether CI-V reports any change in status to the controlling PC program, or not. In the default ON mode, changing the VFO and other status settings is reported to the PC program. In the OFF position, the status data is only sent if the controlling PC program asks for it. Most PC software can only read data that has been specifically requested, so either setting should work fine.

4. USB/LAN→REMOTE is the address that is used for the RS-BA1 remote control software. Wayne Phillips has published a useful video at https://www.youtube.com/watch?v=pV4_xDtsMoY.

5. CI-V USB Port. 'Link to [REMOTE]' causes CI-V commands to be sent to the REMOTE jack as well as the USB cable. 'Unlink from [REMOTE]' separates the Remote jack from the USB one.

6. The CI-V USB Baud Rate sets the data rate to the PC over the USB cable. But only if the CI-V USB Port is set to the default setting of 'Unlink from [REMOTE].'

7. CI-V USB Echo Back: If set to ON, the radio sends all received CI-V commands back to the software as a confirmation that they have been received. This is almost never required and if the PC software is also set to Echo ON, it can lead to problems. However early versions of WSJT-X require Echo to be ON.

8. CI-V DATA Baud Rate is the data rate for the 2.5mm Data Jack used for connection of a GPS receiver and for the DD 128 kB Data mode.

9. CI-V DATA Echo Back is the Echo setting for 2.5mm Data Jack.

➢ Audio Levels (USB cable)

	Setting	IC-9700 default	ZL3DW setting	My Setting
	Windows 10 mixer input level to radio	N/A	28	
	Windows 10 mixer output level from radio	N/A	50	
1	AF Output Level	50%	50%	
2	USB MOD Level	50%	18%	
3	DATA OFF MOD	MIC, ACC	MIC, ACC	
4	DATA MOD	USB	USB	

1.   AF output level is the audio level being sent to the PC over the USB cable. Since you can change the PC soundcard settings and often the digital mode software receive level, I chose to leave the AF output level set at the default setting of 50%. Select <MENU> <SET> <Connectors> <USB AF/IF Output> <AF Output Level>. (Output Select should be left at AF). AF SQL should be left at OFF (Open).

2.   USB MOD level sets the level that will modulate the transmitter. Use the USB MOD control to set the transmit power while sending a digital mode from an external program. The ACC MOD Level control sets the transmit audio level from the ACC 1 jack. Select <MENU> <SET> <Connectors> <MOD Input> < USB MOD Level>.

3.   DATA OFF MOD sets the input source for the SSB, AM, and FM voice modes when DATA is turned off. It should be set to MIC or MIC & ACC. Select <MENU> <SET> <Connectors> <MOD Input> <DATA OFF MOD> <MIC or MIC, ACC>. You can select <MIC, USB> if you want to transmit voice files from a logging program such as N1MM. But, be sure to use the SSB-D digital mode for all digital mode transmissions or you are likely to transmit a distorted signal. **Note also that in the <MIC, USB> mode the microphone will be live while messages are being sent.**

4.   DATA MOD sets the input source for the SSB-D, AM-D, and FM-D data modes when DATA is turned on. It is normally set to USB. But you could use ACC if you are using the ACC jack for audio, or MIC, USB if you are using the microphone connector for audio connections. Select <MENU> <SET> <Connectors> <DATA MOD>.

Function

The function button is the second button below the touch screen. The first screen displays ten Soft Keys for enabling and disabling the; preamplifier, attenuator, AGC setting, manual and auto-notch, noise blanker, noise reduction, IP+, VOX, Compressor, ¼ tuning rate and transmit monitor. Several of these functions can also be turned on or off using front panel buttons or touch screen controls. A blue line around the icon indicates that the function is turned on or enabled. An amber line around the P.AMP/ ATT Soft Key indicates that the attenuator has been selected.

> ➤ **P.AMP/ATT Soft Key**

The P.AMP/ATT Soft Key controls the status of the internal preamplifier and front-end attenuator. The Soft Key has the same function as the front panel P.AMP/ATT button.

Pressing the button turns the Preamp on or off. The preamplifier gain seems to be about 12 dB on the 2 m and 70 cm band and 5.5 dB on the 23 cm band. Touch and hold the P.AMP/ATT Soft Key to enable a 9 dB attenuator. A blue ring around the Soft Key icon indicates the preamplifier is turned on. An orange ring indicates the attenuator is in circuit.

➤ AGC Soft Key

The AGC Soft Key cycles through the AGC fast, mid, and slow settings. Fast AGC is always selected for the FM mode.

Touching and holding the AGC Soft Key opens the AGC setting screen. To change an AGC time constant, select the operating mode SSB, AM, FM, RTTY, or a DATA mode using the normal mode selection icon. You can change radio modes without leaving the setup submenu. Then select FAST, MEDIUM, or SLOW. Use the Main VFO knob to adjust the value, not the Multi knob. Setting the time constant below 0.1 seconds turns the AGC off. Touch and hold the DEF icon to return the settings for the current radio mode to the Icom default. Press the EXIT button to close the window.

I have not changed any of these settings. I am sure that the default settings are appropriate. Unless AGC is off, a blue indicator is always displayed around the AGC Soft Key.

➤ NOTCH function

Touching the Notch icon cycles the notch filter, through auto, manual, and off. The FM mode does not include the manual notch filter option. The CW and RTTY modes do not include the automatic notch filter option because it would notch out the wanted signal and the Notch filter is not available in the DV mode.

The automatic notch filter (AN) and manual notch filter (MN) eliminate the effect of long-term interference signals such as carrier signals that are close to the wanted receiving frequency. Touch and holding the NOTCH Soft Key opens the MULTI menu where you can adjust the manual notch frequency and width. Start with a narrow notch and if it does not completely remove the interference signal, increase the width. The automatic notch filter will find an eliminate fixed carrier signals, "birdies."

➤ NB (Noise Blanker)

Touching the NB Soft Key turns the noise blanker on. The Soft Key has the same function as the NB button on the front panel. The noise blanker is disabled when the radio is in FM or DV mode.

Touch and hold the NB Soft Key to show the MULTI sub-menu where you can adjust the Level, Depth, and Width. The noise blanker is designed to reduce or eliminate regular pulse-type noise such as car ignition noise. You may need to experiment with the controls when tackling a particular noise problem.

The LEVEL control (default 50%) sets the audio level that the filter uses as a threshold. Most DSP noise blankers work by eliminating or modifying noise peaks that are above the average received signal level. They usually have no effect on noise pulses that are below the average speech level. Setting the NB level to an aggressive level may affect audio quality.

The DEPTH control (default 8) sets how much the noise pulse will be attenuated. Too high a setting could cause the speech to be attenuated when a noise spike is attacked by the blanker. This could cause a choppy sound to the audio.

The WIDTH control (default 50%) sets how long after the start of the pulse the output signal will remain attenuated. Set it to the minimum setting that adequately removes the interference. Very sharp short duration spikes will need less time than longer noise spikes such as lightning crashes.

Noise blanking occurs very early in the receiver DSP process. It is performed on the wideband spectrum before any demodulation or other filtering takes place. Noise reduction is performed on the filtered signal i.e. within the receiver passband.

➢   NR (Noise Reduction)

The noise reduction system in the IC-9700 is very effective. Touching the NR Soft Key has the same effect as pressing the NR button. Touch and hold the NR Soft Key to bring up a sub-menu where you can adjust the noise reduction level (default 5). Adjust the level to a point where the noise reduction is effective without affecting the wanted signal quality.

Noise reduction filters are aimed at wideband noise especially on the low bands rather than impulse noise which is managed by the noise blanker. The NR filter works best when the received signals have a good signal to noise ratio. By introducing a very small delay, the DSP noise reduction filter is able to look ahead and modify the digital data streams to remove noise and interference before you hear it. This is not easily achievable with analog circuitry. It would require analog delay lines.

➢   IP+ Soft Key

The IP+ Soft Key turns on the IP+ function. There is no touch and hold function. Turning IP+ on optimizes the sampling system for best receive IMD (intermodulation distortion) performance. It can improve receiver performance in the presence of very large interfering signals. When IP+ mode is selected, a blue indicator is displayed around the IP+ Soft Key.

The IP+ function turns on ADC randomization. While this may help you get good results during two-tone testing, there is normally enough band activity to make the use of ADC randomization unnecessary. In most cases, the IP+ function will have little or no noticeable effect and it can be left turned off.

➢   VOX Soft Key

VOX stands for voice-operated switch. When VOX is on, the radio will transmit when you talk into the microphone without you having to press the PTT button. In voice modes, the VOX Soft Key has the same function as the VOX/BK-IN button.

Using VOX is popular if you are using a headset or a desk microphone. Touching the VOX Soft Key turns the VOX on or off.

Touch and hold the VOX Soft Key to open the VOX setup MULTI submenu. Touch an icon to select the item to be adjusted. Turn the MULTI knob to change the setting. You should take the time to set the VOX up carefully as some settings tend to counteract other settings. In summary,

- VOX GAIN sets the sensitivity of the VOX, i.e. how loud you have to talk to operate it and put the radio into transmit mode. (Default 50%).

- ANTI-VOX stops the VOX triggering on miscellaneous noise like audio from the speaker or background noise. Higher values make the VOX less likely to trigger. (Default 50%).

- DELAY sets the pause time before the radio reverts to receive mode. It needs to be set so that the radio keeps transmitting while you are talking normally but returns to receive in a reasonable time when you have finished talking. (Default 0.2 seconds).

- VOICE DELAY sets the delay after you start talking before the transmitter starts. Generally, you want the radio to transmit immediately to avoid the first syllable or word being missed from the transmission. However, if you are prone to making noises perhaps you should set it longer. If you start each transmission with "Ah" you could set it quite long. (Default OFF).

➢   BKIN (Break-in)

In CW mode the BKIN Soft Key replaces the VOX Soft Key and the BKIN Soft Key has the same function as the VOX/BK-IN button. Break-in operation is covered extensively in the 'Setting up the radio for CW operation' section on page 20.

> COMP Soft Key

The Compressor Soft Key is visible when the radio is set to one of the three 'voice' modes; AM, FM, or SSB. Touching the Soft Key turns the compressor on or off. A blue indicator around the Soft Key indicates that the compressor is on.

Touch and hold COMP to set the compressor level using the MULTI submenu. Touch the COMP icon to select and turn the MULTI knob to change the setting. (Default level = 5). See setting up the radio for SSB transmission, on page 16.

> TONE Soft Key

The TONE Soft Key is only visible when the transceiver is in FM mode. TONE is used for repeater operation primarily on the 6m band. The Soft Key has three modes; OFF, TONE, and TSQL. A blue indicator indicates that tone is in use.

- The TONE mode sends a tone with your FM transmission that opens the repeater squelch.

- The TSQL mode squelches your receiver until the correct tone is received from the repeater. It also sends the tone on your transmission to open the repeater squelch.

- The OFF mode turns off both receiver tone squelch and the transmitted tone.

Touch and hold the TONE key to open the setup screen.

- REPEATER TONE is the tone that is sent to the repeater in the TONE mode.

- T-SQL is the tone that must be received from the repeater to open the receiver squelch in the TSQL mode. The same tone is sent with your transmit signal.

The full set of CTCSS tones is available. Most repeaters use the default tone of 88.5 Hz or 67 Hz.

There is a useful icon called T-SCAN on the TONE FREQUENCY sub-menu.

- To find the tone that you should be transmitting for the TONE mode. Tune your receiver to the repeater input frequency then touch the REPEATER TONE icon. When another station is using the repeater, touch T-SCAN. The radio will scan through all the possible CTCSS codes until it finds the one that matches the tone on the other station's repeater input transmission. This method will probably work if the receiver is tuned to the repeater output frequency, as the same tone is usually transmitted by the repeater.

- To find the tone that you need for tone squelch in the TSQL mode. Tune your receiver to the repeater output frequency then touch the T-SQL - TONE icon. When another station is using the repeater, touch T-SCAN. The radio will scan through all the CTCSS codes until it finds the one that matches the tone on the repeater's output transmission.

Touch and hold DEF to reset the tone back to the default of 88.5 Hz.

See 'Setting the Tone' on page 40 for more information and the list of CTCSS tones.

➢ D.SQL (Digital Squelch)

In the DV mode, the Tone Soft Key is replaced with D.SQL. The default is OFF. D.SQL sets the radio so that the squelch will only open if the received signal includes your call sign. You will not hear any other conversations on the frequency. CSQL is digital code squelch. When CSQL is selected the receiver squelch will only open if the received signal includes the correct code. Touch and hold the Soft Key and use the main VFO tuning knob to select a code between 00 and 99.

➢ TBW (Transmit Bandwidth)

Touch the TBW Soft Key to cycle through the three transmit bandwidth settings, WIDE, MID, or NAR. There is no touch and hold function.

WIDE is suitable for Rag Chewing on the local Net or chatting to locals on the 80m or 10m band. MID is better suited to working DX and the NAR (narrow) mode is suited to contest operation.

If you want to change the actual TBW bandwidth for each of the three options, see 'Setting the transmit bandwidth (TBW),' on page 18.

➢ MONI (Transmit Monitor)

The MONI Soft Key turns on the audio monitor so that you can hear the signal that you are transmitting. Touch and hold the Soft Key to show the MULTI submenu. Turning the Multi knob adjusts the level of the monitor signal. A blue indicator around the Soft Key indicates that the transmit monitor is turned on.

'Monitor' is not available in the CW mode because sidetone is always on.

Unless you specifically want to listen to your transmit signal it is less distracting to leave the Monitor turned off. The SSB voice keyer will be heard if the Auto Monitor setting is 'on,' irrespective of the Monitor setting.

> ¼ tuning speed

The digital modes and CW allow the use of the ¼ tuning function. Select
<FUNCTION> <1/4> to turn the function on or off. The function slows down the
tuning rate of the VFO to make tuning in narrow CW and digital mode signals easier.
It is indicated with a ¼ icon to the right of the 10 Hz digit of the frequency display.
It is indicated above the 1 Hz digit if the 1 Hz tuning step is active. Touch and hold
the 10 Hz digit on the VFO display to switch to the 1 Hz tuning step.

The second Function sub-screen contains three additional settings.

> DUP

DUP turns on the repeater duplex offset. Touch repeatedly to cycle through no
offset, plus offset, and minus offset settings. The offset is usually pre-set for the band,
but you can change the offset by touch and holding the DUP Soft Key.

> EXT P.AMP

This control sets whether the radio will send a DC voltage up the coax feeder cable
to power an external masthead amplifier. The control is intended as a quick way to
turn the external preamplifier on or off.

If you have a masthead preamplifier you must also turn on the DC feed option in
the menu. There is a separate control for each band/antenna connector. <MENU>
<SET> <Connectors>

> RPS

RPS is the 'repeater simplex' mode. It is only available in DD (128 kbit digital data)
mode on the 23 cm band. There is no touch and hold function. There is only one
reference to this mode in the Icom Advanced Manual. It says to turn it off if you are
operating in DD simplex mode without a repeater. I think that it is probably used
when you are operating DD mode through a simplex 'hot-spot' rather than a duplex
repeater.

> TX PWR LIMIT

The TX Power limit is used to limit the maximum power on a particular band. For
example, you may be transmitting on the 2m band into a linear amplifier or
transverter that can only accept 10 Watts. In that case, you should set the TX Power
limit to 10% of maximum power to avoid the possibility of damage to the amplifier
or transverter.

Touch the Soft Key to turn the limit on and then Touch and hold the Soft Key and
use the MULTI knob to set the percentage of full power that you want.

M.SCOPE

➢ Press M.SCOPE

Pressing M.SCOPE toggles between a display with big VFO numbers and a small panadapter display. The small panadapter display has no Soft Keys so you can't change any of the panadapter settings unless you press and hold M.SCOPE to change to the big panadapter

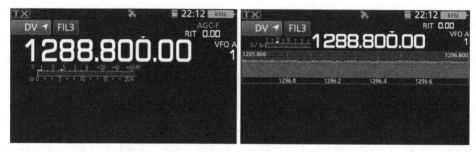

Pressing the M.SCOPE button toggles between these two displays

➢ Press and hold M.SCOPE

Depending on whether the 'EXPD/SET' Soft Key has been selected, holding M.SCOPE down for one second will change the display to either big VFO numbers with a small panadapter, or small VFO numbers with a big panadapter display. Use the 'EXPD/SET' Soft Key to toggle between panadapter sizes

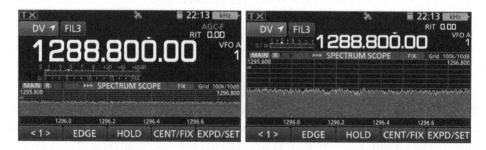

Press and hold M.SCOPE. Then touch EXPD/SET to toggle between these two displays

Quick

The QUICK button is a way of accessing setup and information screens without having to go through the main menu steps. The choices are different if you are in the DR digital repeater mode. The QUICK menu is also used as an intermediate step in some setup functions. In the standard modes the choices are:

➢ VFO/MEMORY

This item opens the same sub-menu as touching the VFO or MEMO icon on the touch screen. Except you have to push the QUICK button first so it is not really quicker.

➢ METER SELECT

Selecting Meter Select opens a two-page sub-menu which lets you select the information that the meter displays while the transceiver is transmitting. This function is exactly the same as touching the meter scale on the display except that you go directly to the choice you want rather than having to cycle through the options.

➢ GPS Information

This is a quick way of displaying the GPS information screen. It is only valid if you have a GPS receiver connected to the radio. It is the same screen as you get through the main menu GPS sub-menu but easier to access.

➢ GPS Position

This is a quick way of displaying the four GPS position screens. It is the same screen as you get through the main menu GPS sub-menu but easier to access.

➢ Home CH Set

This function is not described in either of the Icom manuals. I believe that it sets a frequency on each band that is known as the home channel. When you tune the VFO across the frequency you get a beep.

➢ <<REC Start>> (and <<REC Stop>>)

<<REC Start>> starts recording audio to a file on the SD card. While a recording is in progress pressing QUICK again shows that the <<REC Start>> option has become <<REC Stop>>. Touching that stops the recording.

There are menu settings controlling what is recorded and for how long. See RECORD on page 118.

You can elect to record both your transmissions and the signal received off the air, or only the received signals. You can elect to record every transmission every time the microphone PTT is pressed.

In the DR mode the choices are:

➢ Group Select

Group Select lets you a DV/DD Repeater group. Only groups that are not skipped are visible. That will usually only be your local area. The top channel will be loaded into the 'From' box. With the 'From' box highlighted use the main VFO to step through the un-skipped repeaters.

➢ Repeater Detail

This screen shows the details of the currently selected repeater. The detail screen can also be entered from the FROM SELECT screen. The information provided includes the; repeater name, sub name, group number, call sign, repeater type, frequency, duplex setting, and the direction and distance from your location.

➢ Memory Write

Saves the current DR repeater information to a DV/DD memory slot. touch <Memory Write>, select a blank memory slot (or a slot to overwrite), and touch <YES>.

➢ Meter Select

Selecting Meter Select opens a two-page sub-menu which lets you select the information that the meter displays while the transceiver is transmitting. This function is exactly the same as touching the meter scale on the display except that you go directly to the choice you want rather than having to cycle through the options.

➢ GPS Information

This is a quick way of displaying the GPS information screen. It is only valid if you have a GPS receiver connected to the radio. It is the same screen as you get through the main menu GPS sub-menu but easier to access.

➢ GPS Position

This is a quick way of displaying the four GPS position screens. It is the same screen as you get through the main menu GPS sub-menu but easier to access.

> SKIP

The SKIP setting sets the current repeater channel to be skipped. 'SKIP' will show in the 'From' text box. Press <QUICK> and touch SKIP again to reset.

> Home CH Set

This function is not described in either of the Icom manuals. I believe that it sets a frequency on each band that is known as the home channel. When you tune the VFO across the frequency you get a beep.

> <<REC Start>> (and <<REC Stop>>)

<<REC Start>> starts recording audio to a file on the SD card. While a recording is in progress pressing QUICK again shows that the <<REC Start>> option has become <<REC Stop>>. Touching that stops the recording.

There are menu settings controlling what is recorded and for how long. See RECORD on page 118. You can elect to record both your transmissions and the signal received off the air, or only the received signals. You can elect to record every transmission every time the microphone PTT is pressed.

> >>Normal Mode>>

Cancels the terminal or access point D-Star modes

Exit

The EXIT button can be pressed to exit from any of the sub-menu displays including the MULTI sub-menus.

Rear panel connectors

REAR PANEL CONNECTORS

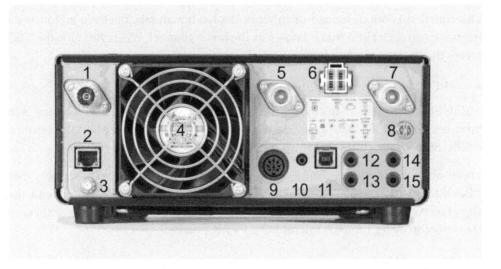

1.	Antenna jack for the 2 m band (SO239 use a 50 Ω PL-259 plug)
2.	Ethernet LAN (RJ45)
3.	10 MHz reference signal input (SMA)
4.	Cooling fan
5.	Antenna jack for the 70 cm band (use a 50 Ω Type-N plug)
6.	DC 13.8 Volt power supply
7.	Antenna jack for the 23 cm band (use a 50 Ω Type-N plug)
8.	Ground Lug
9.	ACC jack (8 pin DIN)
10.	DATA jack (2.5mm 1/10″ stereo mini phono)
11.	USB port (USB 2.0 Type B)
12.	Remote CI-V (3.5mm ¼″ mono phono)
13.	KEY straight key or paddle (3.5mm ¼″ stereo phono)
14.	External speaker - Main receiver (3.5mm ¼″ stereo phono)
15.	External speaker - Sub receiver (3.5mm ¼″ stereo phono)

ANT 1

The SO239 connector is the primary antenna jack used for receiving and transmitting on the 2 m (144 – 148 MHz) band. The connector is a 50 Ω SO-239 jack. It takes a PL259 'UHF' plug.

'UHF' connectors were designed in the 1930s for the US Navy Signal Corps for use with RG213 coax. They can also be fitted to LMR-400 and to RG58 coax using an adapter sleeve. They are commonly known by their original U.S. Navy part numbers, PL259 for the plug and SO239 for the socket.

DC can be sent up the coax cable to power an external preamplifier. <MENU> <SET> <Connectors> <External P.AMP>.

LAN

The LAN port is a standard RJ-45 100T Ethernet port. It has the usual LED indicators which blink when the radio is communicating with the LAN. The LAN port is used for synchronizing the clock to an NTP server and for the Icom RS-BA1 remote control software. Unlike earlier models, you do not have to have a PC at the radio end of a remote-control setup. The remote-control server is built into the radio. The port can also output audio or a 12 kHz IF to the PC and it can connect the radio to the Internet for data transfer using the DD or DV modes. You cannot do a firmware update using the LAN connection.

If you are not intending to use the other LAN functions, I am not sure that it is worth connecting a LAN cable just for the clock synchronization function.

REF IN

The REF IN (SMA) connector allows you to connect an external 10 MHz frequency reference.

Since firmware release V1.10 the external frequency reference can be used to continuously synchronize or govern the stability of the radio's internal oscillator. The external 10 MHz source must be super accurate GPS referenced clock or a Rubidium frequency standard with low phase noise because the radio already has a TXCO with a stability of ± 0.5 ppm. The required input is a 10 MHz, 50 Ω unbalanced, clock signal at around -10 dBm. The native (non-governed) frequency stability

- will be better than ± 74 Hz at a frequency of 148 MHz.
- will be better than ± 225 Hz at a frequency of 450 MHz.
- will be better than ± 650 Hz at a frequency of 1300 MHz.

COOLING FAN

Make sure that the fan is unobstructed.

ANT 2

This Type N connector is the primary antenna jack used for receiving and transmitting on the 70 cm (430-450 MHz) band. Use a 50 Ω Type N plug.

DC can be sent up the coax cable to power an external preamplifier. <MENU> <SET> <Connectors> <External P.AMP>.

DC 13.8V

Connect the radio to a 13.8 Volt (± 15%) regulated power supply capable of supplying at least 18 Amps (20A recommended). Use the supplied power lead.

BE VERY CAREFUL TO SUPPLY THE CORRECT DC POLARITY. Red is positive and black is negative. **Never use a power lead without inline fuses.**

The connector has a locking tab. It is a bit fiddly to use. Squeeze the tab on the top of the connector down before attempting to unplug the DC cable.

ANT 3

This Type N connector is the primary antenna jack used for receiving and transmitting on the 23 cm (1240-1300 MHz) band. Use a 50 Ω Type N plug.

DC can be sent up the coax cable to power an external preamplifier. <MENU> <SET> <Connectors> <External P.AMP>.

Some people have asked why the radio does not have SO239 connectors for the 70 cm and 23 cm bands. Despite the name, 'UHF' connectors are not actually usable at UHF frequencies (above 300 MHz). Nearly all SO239 sockets have a non-constant characteristic impedance resulting in increased mismatch loss at higher frequencies. They are only suitable for use up to about 160 MHz. In the 1930s, when they were designed, anything over 30 MHz was considered to be an 'ultra-high' frequency. They should have been renamed 'HF' connectors, but most people refer to them by their old U.S. Navy part numbers, PL259 for the plug and SO239 for the socket.

GND

The radio ground connection should be connected to your shack ground in a 'star' rather than daisy chain format. The shack ground should be connected to an earth stake or earth mat outside. NOT to the mains earth. Earthing the radio can protect the radio from lightning static discharge (not lightning strikes). It can also improve noise performance.

ACC

ACC (accessory jack) is an 8 pin DIN connector. It is intended for audio and control from a PC via an interface box. Something like an MFJ or RIGblaster interface. Other

You are more likely to use the USB connector which can do these functions with one cable. You might perhaps use an external interface box if you are operating in a multi-radio setup. The interface box should have audio transformers to isolate the receive and send audio signals and optocouplers or transistors to drive the RS232 control lines. See the Icom Basic Manual page 13-1 for more details.

Pin	Name	Function
1	RTTY	FSK Keying for RTTY (FSK mode only) 'high' > 2.4 V, 'low' ground i.e. < 0.6 V, output current < 2 mA
2	GND	Ground (connect to a 'telecommunications' or 'station' earth rod, not the Mains earth).
3	SEND	Send / PTT input and output (goes low on transmit). If you are switching a relay, always put a reverse-biased protection diode across the relay coil to protect the radio from back EMF. Input 'high' = 2-20V, 'low' = ground i.e. -0.5 to +0.8V. Output 'low' = ground i.e. less than 0.1V max current 200mA.
4	MOD	Audio input to the radio, impedance 10 kΩ, level 100 mV
5	AF/IF	Audio, or 12 kHz IF output, from the radio, (menu choice). Level 100-300 mV at 4.7 kΩ
6	SQL S	Squelch output (goes low when the squelch opens). 'high' squelch closed >6.0 V 100 uA 'low' squelch open Ground i.e. < 0.3 V max 5 mA
7	13.8V	13.8 Volts at 1 Amp maximum
8	ALC	ALC input 0 to -4 Volts. Input impedance >3.3kΩ

DATA

The Data jack is a 2.5mm (1/10") stereo mini phono jack. Transmit data TxD is on the connector tip and receive data RxD is on the ring. The sleeve is common ground.

The Data jack can be used for DV gateway connection or remote control via CI-V commands using the OPC-2350LU cable and the RS-MS3W (Windows) or RS-MS3A (Android) software. However, it is easier to use the LAN port connection.

The most useful function of the Data jack is for connection to a NMEA format compatible external GPS receiver. The cable from the Data jack can be directly connected to an RS-232 connector plugged into the GPS receiver. RS-232 pin 5 is GND, pin 2 is RxD and pin 3 is TxD. The Icom manual states that USB to RS-232 adapters are not supported, but if your GPS receiver only has a USB interface, some models will probably work OK. Your GPS position, altitude, and speed can be displayed on the IC-9700 display, and the data can be added to DV and DD mode transmissions.

USB

The USB port is a USB 2.0, Type B port. You will need to purchase a 'Type A to Type B USB 2.0' cable to connect the radio to a PC. You will also have to download and install the USB driver software. See page 23.

USB is used for CI-V CAT control of the transceiver and for transferring audio to and from a PC for external digital mode software. It is also used to send decoded RTTY data to the PC and for remote control via the Icom RS-BA1 software. Using the CS-9700 software you can clone radio settings to or from another IC-9700.

For digital mode operation, you may have to set up the CI-V controls and the RTS and DTR keying lines, (page 137).

KEY

The rear panel KEY jack is for the connection of a CW bug, paddle, or straight key. Select CW mode. <MENU> <KEYER> <EDIT/SET> <CW-KEY SET> <Key Type> and select straight, paddle or bug.

The Key jack is a non-standard, 1/8" (3.5 mm) stereo phono plug. The wiring is standard. Dit on the phono plug tip, Dah on the ring, and common on the sleeve. For a straight key use the ring and the sleeve connections only.

If you want to use a straight key, bug, or external hardware keyer you will have to turn off the internal keyer. Select CW mode. <MENU> <KEYER> <EDIT/SET> <CW-KEY SET> <Key Type> and select 'Straight.'

One setting that is not included in the KEYER EDIT/SET menu structure is the choice of having the CW signal on the transmitter's upper sideband or lower sideband. That is in <MENU> <SET> <Function> <CW Normal Side>. The default is USB.

REMOTE

The REMOTE mono mini phone jack is used to connect the radio to a PC via an Icom CI-V to RS232 or USB adapter. It is there for compatibility with older radios and peripherals. If possible, use a USB cable connected to the USB port instead.

MAIN RECEIVER SPEAKER JACK

The speaker jack on the rear of the radio is for connection to an external speaker. It is a 1/8" (3.5mm) mono jack with a mono output ⊗.

There is a menu setting to send a mix of both the sub and main receiver audio to the speaker jack, or just the main receiver output. You can't have the main receiver sound from the internal speaker and the rear speaker jack at the same time.

SUB RECEIVER SPEAKER JACK

The speaker jack on the rear of the radio is for connection to an external speaker. It is a 1/8" (3.5mm) mono jack with a mono output ⊗.

There is a menu setting to send a mix of both the sub and main receiver audio to the speaker jack, or just the sub receiver output. You can't have sub receiver sound from the internal speaker and the rear speaker jack at the same time.

D-Star

The Icom Advanced Manual devotes four chapters to D-Star operation. A total of eighty-one pages, compared to the section on Satellite operation which only rates three pages. I am completely new to D-Star as this is my first radio with D-Star capability. So, I will cover the basics and some operating tips for D-Star operation in this chapter and leave the heavy technical details to Icom.

D-Star (Digital Smart Technologies for Amateur Radio) is an open-source digital voice and data protocol for amateur radio. The system was developed in the late 1990s by the Japan Amateur Radio League with support from Icom. It uses less RF bandwidth than FM and provides completely noise-free audio quality. You can use the D-Star 'DV' (digital voice) mode on all of the bands that the IC-9700 supports. The 23 cm band also supports the DD (digital data) mode which can be used to transfer data files at up to 128 kbits per second and even as a remote link to the Internet for a connected PC.

You can carry out simplex conversations between your radio and other D-Star capable radios, or you can talk through a D-Star repeater to D-Star capable stations in your area. But the major advantage of D-Star and the main reason for its popularity is the ability to use D-Star or mixed-mode 'Gateways' and 'Reflectors.'

Gateway operation allows you to create a link via the Internet between your local repeater and almost any other D-Star repeater worldwide. Once the link is established you can put out a CQ call, respond to others using the distant repeater, or call a station that is within the coverage of the distant repeater. It should be noted that any call you make to another station is not private. It will probably be broadcast by both your local repeater and the connected repeater. After you have finished your call, if you created the link you should send the 'unlink' signal to remove the linking between repeaters and you should probably also re-instate any link that was in effect before you started. Icom states in the Advanced Manual that the radio comes with a comprehensive list of D-Star repeaters already loaded. The radio can store 2500 memories for D-Star repeaters, in addition to the normal 99 channels. My radio was supplied with no repeaters loaded at all, but that is easily rectified. See the section on loading D-Star memories on page 160.

The DV/DD memory channels can hold data for FM repeaters as well as D-Star repeaters and you can use the DR (digital repeater) function to access them. It is also possible to store DV frequencies in the standard 99 channel memory banks. You can save individual channels for linking to specific repeaters or reflectors, but overall it is not as easy and flexible as using the DR mode. I believe that it is better to put all your DV repeaters into the DV/DD memory slots and your FM repeaters into the standard memory slots.

Reflector operation allows you to create a link via the Internet between your local repeater and a 'Reflector.' A reflector is a way of linking your repeater to a group of other connected repeaters. Reflectors are often used for special interest group 'Nets.' You cannot choose which other repeaters are connected to a reflector. Users within each repeater coverage area can connect or disconnect their repeater as they wish. Once the link is established you can put out a CQ call, respond to others using the distant repeaters, or call a station that is within the coverage of one of the connected repeaters. As with linked gateways, any call you make to another station is not private. It will usually be broadcast on your local repeater and all of the other connected repeaters worldwide. If you established the connection to a reflector, you should send the 'unlink' signal to remove the link after you have finished making calls. You should probably also re-instate any link that was in effect before you started. If you didn't establish the connection, and you don't want to make a different connection, you should leave the link connected. The rules may be different for your local repeaters. The radio does not have memory slots for Reflectors, but it does store the last few that you have used.

FIRST STEPS

The very first thing that you have to do is enter your callsign into the radio. No digital mode conversations can be carried out unless this simple step is done. Use <MENU> <1> <My Station> <My Call Sign (DV)>. Touch and hold the top position and then touch 'Edit.' Enter your callsign before the / symbol. After the / symbol you can enter a four-digit code. For the IC-9700 you would normally enter 9700. But it could be P for portable, M for mobile, VK if you are temporarily operating in Australia, or K4 if you are temporarily operating in the K4 call area. Don't put your name after the / symbol. There is a better place for that.

You can enter up to eight callsigns. Perhaps more than one operator will use your radio, or you may want to add separate entries for /9700, /P, /M. Most people will only use one callsign memory slot.

Press ENT to save the changes or they will be lost. Then exit using the ↺ Soft Key twice or the EXIT button twice to get back to the 'My Station' setup screen.

You may notice that the 'My Call Sign (DD)' entry has updated to show the same callsign as you entered for the DV mode. You can change it if you want to.

On the next line 'TX Message (DV)' you can enter some information that will be displayed when other stations see and hear your transmission. Most people put their name and location or their Maidenhead grid reference. <MENU> <1> <My Station> < TX Message (DV)>.

After your callsign has been entered. It will be displayed on the Icom splash screen when you turn on the radio.

REGISTRATION

You can make local repeater calls and point to point simplex calls to another D-Star radio without any further complications. But if you want to talk over a linked repeater or reflector you must register your callsign with the D-Star organization. It is an online process done from your Internet web browser, rather like registering for an Internet forum. Ask someone how it is done in your area or do a Google search based on the repeater callsign and 'registration.' Most people register to a repeater near them because it becomes your 'home repeater.' Your home repeater is your default location. If someone places a call to your callsign the system will send the call to the repeater that you last used. But if it can't find you listed, it will link to your home repeater and put the call out there.

Usually, registration is a two-stage process, but some repeater moderators have automated the process. You enter your callsign and create a password. In a day or two, you should get an email from the repeater moderator. Armed with the required information you go back to the website and enter a small amount of additional information. Check the checkbox on the first line. Enter your callsign in uppercase (it may be pre-populated). In the box marked 'Initial' enter a space. You **must** enter a single space. Leaving the field blank will not work. In the 'pcname' field enter your callsign in lower case. This **must** be in lowercase.

You will notice a 'local IP' address has been generated, but you don't need to remember it. You won't need it again. If you operate two D-Star radios, for example, a mobile and a home station, you fill in two lines, one for each D-Star radio that you are registering. You would also complete an additional line if you were planning to use the IC-9700 as a gateway or a repeater ('access node'). You can always do it later.

Press the 'Update' button to complete the process.

Congratulations! You will never have to do this again unless you move permanently to a new location or you need to register another radio or a hot-spot.

DV/DD MEMORY

The DV/DD memory 'Repeater List' is used to store D-Star digital voice and data channels for all three bands. It can hold 2500 records divided into 50 repeater groups. This seems like a lot, but the radio can access D-Star and other digital voice repeaters worldwide through local repeaters or hotspots which are linked to worldwide networks via the Internet. Obviously adding all of these channels by hand would be a major hurdle, so the radio has the capacity to load the bank from a CSV file stored on the SD card. You can also use the CS-9700 or RT Systems programmers. Copy the file onto the IC-9700\CSV\RptList directory of the SD card and then import the file using <MENU> <SET> <SD Card> <IMPORT/EXPORT> <IMPORT> or use the IC-9700 or RT Systems programmer.

The Icom manual states that the DV/DD memory bank is pre-programmed, but my radio was supplied blank. You can download a file containing the D-Star repeaters from the download area of the Icom Japan website.

http://www.icom.co.jp/world/support/download/firm/IC-9700/repeater_list/.

I modified the data file to put the New Zealand repeaters into Group 01 so that they appear at the top when I want to use them. I also put my local D-Star repeaters at the top of the Group 01 list. You can include normal FM repeaters in the list if you want to.

The DV/DD memory sub-menu is also where you set the name and call sign of your buddies that you will call a lot using the D-Star mode. <MENU> <DV/DD memory> <Your Call Sign>. Yes, the title "Your Call Sign" is a bit misleading. Perhaps it means 'your list of commonly used call signs?' Touch and hold the highest available entry line to add or edit a call sign and name.

If you touch and hold a memory entry, you can add, edit, move, or delete DV/DD memories. Or skip them during the DR scan. The 'Skip all ON' setting skips the entire group. 'Skip all OFF' includes the entire group in the DR scan.

If you touch and hold a memory group title, you can edit its name or alter the skip setting to skip the entire group, or to include the entire group. There is no point scanning for repeaters in different countries or out of the range of your radio. The only channels that you want to be included in the DR scan are your local repeaters. I found the easiest thing to do was to set 'Skip all ON' on all of the groups and then use the individual SKIP Soft Key to set my four local repeaters so they will not be skipped. In DR mode, press the SCAN button to scan the FROM repeater through your local channels. They can include FM repeaters as well as D-Star repeaters.

If you hear a signal from a linked repeater or gateway that is not in the memory list. You can add the repeater to the list of repeaters in the DD/DV memory from the RX History screen.

CS-9700 MANAGER

The free CS-9700 memory manager from Icom can read memories and menu settings from the radio. You can edit the memory banks in the PC software and then upload them back to the radio over the USB cable.

The program uses the 'clone' mode over the USB interface. You can set most (maybe all) of the menu settings and you can store a copy of the radio configuration on your PC.

I always download a fresh copy of the radio configuration from the radio, make any changes that I want using the software, store the resulting file to the PC, and then upload the revised configuration back to the radio.

MAKING A SIMPLEX CALL

You don't have to use the DR (digital repeater) mode to make a simplex call to another D-Star radio. Just tune the radio to the frequency that you want to use, select DV mode, and make your call. You do not have to be registered to use DV for simplex calls.

MAKING A CALL ON A LOCAL REPEATER (NON-DR MODE)

You don't have to use the DR (digital repeater) mode to make a call over your local D-Star repeater to another local D-Star radio. You can save the repeater information in one of the 99 standard memory channels in the same way that you would save an FM repeater channel.

Select the channel or the repeater output frequency, select the correct Duplex offset (usually automatic), select DV mode, and make your call. You do not have to be registered to use DV for local calls.

I don't recommend using this method unless the repeater is not connected to the Internet and can't be used as a gateway. Use the DR mode instead.

It is very important to note that in this local repeater mode you can only talk to local D-Star equipped stations. You will not be able to reach any station that is arriving through a link to another repeater or reflector. This can be very confusing. You can hear them, but they can't hear you. Sometimes you can hear a call sign that you know is a local operator, but they may still be accessing the repeater through an Internet connection. They may be using a D-Star 'hotspot,' or they may be using a DMR or P25 radio into a repeater that is linked to your D-Star repeater.

Generally, I think that it is best to treat all D-Star repeaters as gateways and always use the DR (digital repeater) mode.

DR (DIGITAL REPEATER) MODE

Press and hold the CALL/DR button to enter the DR (digital repeater) mode. You must be in this mode if you want to use linked 'gateway' repeaters and reflectors.

DR mode also enables you to select from the DV/DD repeater memories, received call signs, and previous contacts.

Next, to the 'To' icon, you will see a text box which contains the UR data, usually 'CQCQCQ.' Next to the 'From' icon, there is a text box which contains the callsign of the repeater that you are using. To the left you will see the red TX indicator, a blue DV indicator (which may flash to FM at times if the band is noisy), and a FIL filter indicator, probably showing that the receiver filter is set to FIL3 which indicates that the receiver is set for a 7 kHz bandwidth, ideal for D-Star.

There is a small "Gotcha" relating to the selection of the repeater that you are planning to use. The IC-9700 has two receivers but they can't be set to receive on the same band at the same time. For example, if you had the Main receiver set to the DR mode on the 2m band and the Sub receiver set to an FM repeater on the 70 cm band. When you touch the 'FROM' icon on the Main receiver, you will not be able to set it to any 70 cm repeaters because the Sub receiver is already using the 70 cm receiver.

You can put both receivers into DR mode and listen to DV repeaters on two different bands at the same time. Or you can have one band on DV and the other on FM. But you cannot listen to two repeaters on the same band at the same time.

The 'To' box will normally show 'CQCQCQ' with an icon showing three little people, or it may show 'Use Reflector CQCQCQ' with an icon showing a computer and screen. These indications are **very important** when you make a call.

➤ Local CQ

Touch the text beside the 'To' icon (possibly twice) to open the 'TO SELECT' sub-menu. If you select 'Local CQ,' the text next to the 'To' icon will show 'CQCQCQ' and the 'To' box will be an icon showing three little people. In this mode, you can only transmit through your local repeater to D-Star stations within the repeater's coverage area. You can't talk to anyone accessing your repeater through a gateway or reflector. You can hear them, but they can't hear you. Generally, unless the repeater is unlinked it is best to always use the 'Use Reflector' mode.

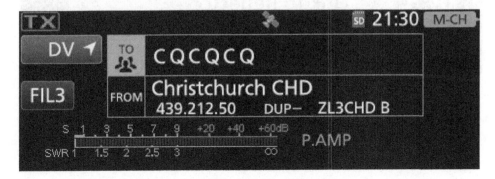

➤ Use Reflector

Touch the text beside the 'To' icon (possibly twice) to open the 'TO SELECT' sub-menu. Touch <Reflector> <Use Reflector>. The text next to the 'To' box will show 'Use Reflector CQCQCQ' with an icon showing a computer and screen. In this mode, you will transmit through your local repeater to D-Star stations within the repeater's coverage area and you can also talk to anyone accessing your repeater through a gateway or reflector. Use this mode whenever the repeater is linked.

USEFUL TIP: Anytime you are DR mode you can change the 'From' settings using the main VFO knob. Touch the 'From' icon and rotate the VFO knob. It will cycle through the repeaters on your repeater list. Similarly, you can change the 'To' settings using the main VFO knob. Touch the 'To' icon and rotate the VFO knob. It will cycle through the repeaters on your repeater list.

USEFUL TIP: When you are in the 'Use Reflector' mode you can change Reflector settings using the main VFO knob. Touch the 'To' icon with the computer on it and rotate the VFO knob. It will cycle through; Use Reflector CQCQCQ, Link to Reflector, Unlink Reflector, Echo Test, and Repeater Information.

D-STAR FIELD STRUCTURE

D-Star uses four special data fields. They are called; UR, R1, R2, and MY. You can see the contents of these four data fields if you select <MENU> <2> <CS>.

➤ UR, 'Your Call' holds the information being sent to the repeater and the Internet. Normally it will be set to 'CQCQCQ.' This could have been labeled 'TALK,' but I guess CQ is more universally understood. The data in the UR field will change if you are sending a signal to obtain information, call a particular call sign, or link to a repeater or reflector. In the table below '_' indicates a space.

- CS Indication is the information that you will see on the CS menu screen <MENU> <2> <CS>.

- DR Indication is the information that you will see in the 'To' box on the DR screen.

CS Indication	DR Indication	Function
CQCQCQ	CQCQCQ	Talk on a local repeater
CQCQCQ	Use Reflector CQCQCQ	Talk on a local repeater or through a gateway to a reflector or a linked repeater
_____I	Repeater Information I	Ask the repeater for information and check if the repeater is linked.
_____U	U	Unlink the repeater so that you can link to a different repeater or reflector, or make a local call without the world listening.
_____E	Echo Test E	Echo can be used on some repeaters. It will send back a few seconds of your own transmitted audio, as a test signal.
ZL2RO	ZL2RO	Call a particular station for example, ZL2RO
/VK8RADC	Darwin VK8RAD C	Gateway connection to VK8RAD 2 m repeater in Darwin Australia
REF001CL	Link to Reflector REF001CL	Connect to the REF001C reflector. The L indicates a reflector.

➢ R1, 'Repeater 1' is set to the call sign of the repeater that you are calling through. The call sign is followed by a letter which must be the eighth character. For example, 'ZL3DV_ _ B' would be a 70 cm repeater in New Zealand.

Module	Function
A	23 cm repeater
B	70 cm repeater
C	2 m repeater
D	Digital Data

If R1 is blank or (NOT USED*) it means that you are making a Simplex Call. No local repeater is selected.

TECHNICAL TIP: Japan uses A for 70 cm repeaters and B for 23 cm repeaters. Some systems have multiple modules on the same band, so B, C, and D might all be 70 cm repeaters. It does not really matter so long as you have the correct letter code for the repeater you will be accessing with your radio. If the repeater call sign is correct but the letter is incorrect you will not get a message back when you transmit. In that case, try a different letter.

➢ R2, 'Repeater 2' is set to the call sign of the gateway. The call sign is always followed by the letter G, for gateway. The G must be the eighth character. For example, 'ZL3DV_ _ G' would be the gateway for the ZL3DV repeater.

Module	Function
G	Gateway

If R2 is blank or (NOT USED*) it means that you are making a Local Repeater Call. No gateway repeater is selected.

➢ MY, 'My Call' holds your call sign including the /9700 suffix that you set in the First Steps above. The field is eight characters before the / symbol and four characters after the / symbol. In the CS sub-menu or the TO SELECT sub-menu you can change from one version of your callsign to another, or edit the data that will be sent. See First Steps above.

MONITORING A DV REPEATER (DR MODE)

Put the radio into DR mode by press and holding the CALL/DR button. To avoid confusion, at least initially, turn off the other receiver if it is running by pressing the bottom AF gain (volume) knob. The 'To' box will probably show 'CQCQCQ' or 'Use Reflector CQCQCQ.' It does not matter if you are not transmitting.

To be on the safe side, if the text does not include CQCQCQ, touch the 'To' text box and then select 'Local CQ.' Then exit using the ⤺ Soft Key.

Next, you have to set the radio to receive signals from a repeater near you. This could be a D-Star or multi-mode digital repeater at a high site, or it might be a 'simplex repeater,' or a high-power hotspot, or a low power D-Star dongle or hotspot connected to a PC very close to you. Touch the text beside the 'From' icon. You may have to touch it twice, to bring up the 'FROM SELECT' sub-menu. You will see three menu options.

- Repeater List – lets you select a repeater from your DV/DD repeater list. Note that it must be a repeater that is near to you. If you are not within the repeater's RF coverage area it won't work.

  If you have no repeaters listed (like me), or your local repeater is not listed, see the section about D-Star memories on page 160.

- Near Repeater - lets you select from a list of repeaters that are located near you. This will only work if you have GPS location data loaded. See the section about GPS on page 180.

- TX History – lets you select from a list of recently accessed repeaters. Note that recently accessed FM repeaters will not be included unless you were using the DR mode. If you are regularly accessing two or three local repeaters, the TX History (DV) option is probably the easiest way to switch between them.

After you have selected your local repeater you should have some information in the 'From' text box and you will hear anyone using the repeater… probably.

There is a potential "Gotcha" here! You will only hear D-Star traffic on the repeater. Some repeaters are capable of multi-mode operation. If you see an indication on the S meter but don't hear any audio, the chances are that a user is using the repeater for DMR, P25, Fusion or another digital transmission mode. there are also private squelch options such as D.SQL and CSQL, but you should not have those active at this stage.

The 'From' box will display the name of the repeater, it's frequency, duplex setting (DUP+ or DUP-) and the callsign of the repeater, with the A, B, or C letter indicating the band. The frequency will change to indicate your transmit frequency while you are transmitting.

If you select an FM repeater (stored in the DV/DD memory list), the radio will switch to FM and the text will display the same information without the repeater callsign.

TECHNICAL TIP: A simplex repeater or hotspot will also show (DUP+ or DUP-) because the radio does not understand the concept of simplex repeaters. Don't be concerned, the duplex offset will be set to 0.000 so the transceiver will transmit and receive on the same frequency.

CHECK REPEATER LINK STATUS

Before you unlink a repeater, it is a good idea to find out if, and where, it is linked. There are four ways to find this out. The repeater will occasionally announce the link connection, either by a message scrolling across the screen or by a voice announcement. But you probably don't want to wait for that to happen. You can go onto the repeater's dashboard page on the Internet and see what linking is in place. Or you can transmit a code to the repeater which will respond with the information.

After you have received the information, return to CQCQCQ mode. Don't leave the radio in Information mode or every call you make will create an Information report from the repeater.

➢ Information using the I command

Touch the 'To' text box. You may have to touch it twice.

Select <Reflector> <Down Arrow> <Repeater Information>. The 'To' box will change to 'Repeater Information I.'

Key the Microphone PTT for about a second. You do not have to wait for the text to scroll.

You will see a message scroll across the screen and in most cases a voice message saying that the repeater is linked to a repeater or reflector or that it is not linked.

➢ Information using the DTMF 0 command

The DTMF codes are primarily for people using handheld radios where a DTMF code can be sent with a single button press. Unless you are using a microphone with a DTMF keypad, it is a little more difficult on the IC-9700.

Press <MENU> <2> <DTMF> <SEND> <Direct Input> <0> <TX>.

To make this easier, you can store commonly used DTMF codes using the EDIT key. That will shorten the sequence to <MENU> <2> <DTMF> <SEND> <d0:0>. See DTMF codes on page 175.

UNLINK A REPEATER

Before you unlink the repeater, put a call out and let the other repeater users know that you are planning to unlink the repeater. Someone may be waiting for a call from a linked repeater or for a Net to start on a reflector. It is good etiquette to restore the linking after you have finished using the repeater. So, check the repeater dashboard on your PC to see where the repeater is linked. Or send an Information request on the radio and take note of the linking information.

To unlink a repeater, you can use the menu structure in the radio to send the U command, or on many repeaters, you can send a DTMF command.

After you have unlinked the repeater, return to CQCQCQ mode. Don't leave the radio in Unlink mode or every call you make will create an Unlink report from the repeater.

➢ Unlinking using the U command.

Touch the 'To' text box. You may have to touch it twice.

Select <Reflector> <Unlink Reflector>. The 'To' box will change to 'Unlink Reflector U.' You can use this command to unlink from a gateway repeater as well.

Key the Microphone PTT for about a second. You do not have to wait for the text to scroll.

You will see a message scroll across the screen and in most cases a voice message saying that the repeater is now unlinked.

Now that you have unlinked the repeater, you can link to another repeater or reflector or you can make a call out to the local area. Touch the 'To' box again and select 'Local CQ' if you want to make a local call, 'Gateway CQ' if you want to select a link to a repeater, or 'Reflector' if you want to connect to a reflector. These functions are covered in the next few sections.

➢ Unlinking using the # DTMF command.

The DTMF codes are primarily for people using handheld radios where a DTMF code can be sent with a single button press. Unless you are using a microphone with a DTMF keypad, it is a little more difficult on the IC-9700.

Press <MENU> <2> <DTMF> <SEND> <Direct Input> <#> <TX>.

To make this easier, you can store commonly used DTMF codes using the EDIT key. That will shorten the sequence to <MENU> <2> <DTMF> <SEND> <d1:#>. See DTMF codes on page 175.

TIP: Sending a ** DTMF code will reset most repeaters to their default link. This is a handy way of setting the repeater back to "normal." <MENU> <2> <DTMF> <SEND> <d2:**>.

THE ECHO FUNCTION

The Echo function works on some repeaters and older reflectors. Using it is not recommended because it can be annoying to other users. Remember that a reflector may have many repeaters connected to it worldwide. The function is not enabled on most new repeaters. It records and repeats back a few seconds of your transmission so that you can confirm that you are connected. But you can usually tell that by looking at the repeater dashboard or sending an Information request.

After you have carried out an Echo test, return to CQCQCQ mode. Don't leave the radio in Echo mode or every call you make will be repeated back to you.

> Echo using the E command

Touch the To text box. You may have to touch it twice.

Select <Reflector> <Echo test>. The 'To' box will change to 'Echo Test E.'

Key the Microphone PTT for about a second and make a short 5-second announcement such as "K3BLAH testing on Repeater K3AA B." You do not have to wait for the text to scroll. If the function is enabled, you will hear your transmission repeated back to you.

After you have made the Echo call you must always touch the 'To' box again and select 'Local CQ' if you want to make a local call, 'Gateway CQ' if you want to select a link to a repeater, or 'Reflector' if you want to connect to a reflector. These functions are covered in the next few sections.

MAKING A LOCAL CALL (DR MODE)

If you want to make a call to your local area and you don't want the world listening, you can use the Local CQ mode. Make sure that the repeater is not linked to another repeater or a reflector. Before you unlink the repeater, put a call out and let the other repeater users know that you are planning to unlink the repeater. Someone may be waiting for a call from a linked repeater or for a Net to start on a reflector. It is good etiquette to restore the linking after you have finished using the repeater. So, check the repeater dashboard on your PC to see where the repeater is linked, or send an Information request on the radio and take note of the linking information.

Touch the text beside the 'To' icon (possibly twice) to open the 'TO SELECT' sub-menu. If you select 'Local CQ' the text next to the 'TO' box will show 'CQCQCQ' and the 'To' icon will include three little people. Then make a standard CQ call or call another station the way you would on an FM repeater.

In this mode, you can only transmit through your local repeater to D-Star stations within the repeater's coverage area. You cannot talk to anyone accessing your repeater through a gateway or reflector. You can hear them, but they can't hear you. Generally, unless the repeater is unlinked it is best to use the 'Use Reflector' mode. See the next section.

MAKING A CQ CALL ON A LINKED REPEATER

If you don't know if a repeater is linked, it is always best to use this method. Touch the text beside the 'To' icon (possibly twice) to open the 'TO SELECT' sub-menu. Touch <Reflector> <Use Reflector>. The text next to the 'To' box will show 'Use Reflector CQCQCQ' with an icon showing a computer and screen. In this mode, you will transmit through your local repeater to D-Star stations within the repeater's coverage area and you can also talk to anyone accessing your repeater through a gateway or reflector. Use this mode whenever the repeater is linked.

Note that if the repeater is a hotspot or simplex receiver, you always have to use this 'Use Reflector CQCQCQ' mode. The Local CQ mode will not work at all on a simplex repeater.

MULTI-MODE DIGITAL REPEATERS

Some repeaters are capable of transponding multiple digital modes, often D-Star, P25, DMR, and Fusion and sometimes FM as well. Special rules apply to the use of these repeaters because users of other equipment will not be able to hear your transmissions and you will not be able to hear theirs. You need to be very aware of the status of the repeater. If you are at home or have an internet connection this is usually done by observing the repeaters 'dashboard' website. The dashboard will show who is using the repeater and the mode that is currently in use.

If the repeater is being used for a different digital mode, you will be able to see a signal on the radio S meter, but you won't be able to hear it. Usually, there is a timer to stop other signals interfering with an ongoing QSO. You can't use the repeater for a different digital mode until the repeater has been clear for a few seconds.

If a D-Star or multi-mode repeater is linked to another repeater or to a reflector, you can make calls to stations that are listening or calling via the distant repeater irrespective of the digital mode that they are using. For example, if you transmit a D-Star signal into your local repeater and it is linked either directly, or through a reflector, to a DMR repeater you can talk to somebody who is using a DMR radio.

USING A SIMPLEX REPEATER

If the repeater you are using is a hotspot or simplex receiver, you have to use the 'Use Reflector CQCQCQ' mode. The 'Local CQ' mode will not work on a simplex repeater and your transmission will not be broadcast locally. The Simplex repeater must always be linked. In this mode, you will transmit through the gateway to anyone accessing your repeater through a gateway or reflector. But you cannot talk to local stations unless they are using a different repeater that is linked in via the gateway.

Touch the text beside the 'To' icon (possibly twice) to open the 'TO SELECT' sub-menu. Touch <Reflector> <Use Reflector>. The text next to the 'To' box will show 'Use Reflector CQCQCQ' with an icon showing a computer and screen.

CONNECTING TO A REPEATER (GATEWAY)

Check the repeater linking status and then unlink the repeater. Following the instructions above. The 'From' box should be set to your local repeater. Touch the text beside the 'To' icon (possibly twice) to open the 'TO SELECT' sub-menu. Touch 'Gateway CQ' then select the Group that the repeater is in. Scroll down the list until you find the repeater you want to connect to. Touch the repeater row. You will be returned to the main DR screen and the repeater you have selected will be in the 'To' text box.

If the repeater is not in the DV/DD memory bank you can use 'Direct Input (RPT)' to enter its callsign. The call sign is followed by a letter which must be the eighth character. For example, 'ZL3DV_ _ B' would be a 70 cm repeater in New Zealand.

Transmit for a second or two. You don't have to wait until the whole message scrolls. This will link your repeater to the target repeater. You should get a scrolled message and probably a voice message to say that the repeater is connected.

Listen for traffic. Someone may be using the target repeater and they won't want you crashing in. If you want to call CQ, select <Reflector> <Use Reflector> to place 'Use Reflector CQCQCQ' into the 'To' text box. If you want to call a specific station you can set the 'To' text box to their callsign using the 'Your Call Sign' list of your buddies, or you can look in the RX History for their callsign. If the callsign is not in either list, select 'Direct Input (UR)' and type in the wanted station's callsign. Then touch ENT.

Transmit and make a CQ call or call the callsign that you want to talk to in the same way that you would on an FM repeater. Transmitting with a call sign in the 'To' text box will send an alert to the wanted station. But all stations on the distant repeater can hear the call.

If your local repeater is a duplex repeater the locals will hear your call as well. When you have made contact with the wanted station, select <Reflector> <Use Reflector> to place 'Use Reflector CQCQCQ' into the 'To' text box. Otherwise, the other station will get a voice and text scrolling alert every time you transmit.

D-STAR REFLECTORS

The radio supports direct linking to D-Star reflectors that have REF or XRF numbers. Since firmware v1.20 you can also connect to multi-mode reflectors using DCS or XLX numbers. Some reflectors use multiple codes. For example, connecting to DCS299 is the same as connecting to XLX299. See 'connecting to a Reflector' below. Some reflectors have permanent connections to modules for other digital modes. This can allow you to talk to DMR, P25, or System Fusion users.

➢ REF

REF is the DPLUS reflector system. It is a closed source proprietary system developed by Robin Cutshaw AA4RC. REF reflectors were the first generation of D-STAR reflectors and the standard is still very much in use. There is a list of REF reflectors at http://www.dstarinfo.com/Reflectors.aspx.

- REF001C in London is D-STAR's "Mega Reflector." It usually has many repeaters connected to it.

- REF030C in Atlanta is also very popular.

➢ XRF

XRF is the Dextra X-Reflector system, originally created by Scott Lawson, KI4KLF. It is the second generation of D-STAR reflectors, and it is open source. A list of XRF reflectors is available at http://www.xrefl.net/.

➢ DCS

DCS is the 'Digital Call Server' reflector system. It is a closed system developed by Torsten Schultze, DG1HT, and now run by Stefan, DL1BH, Peter, DG9FFM, and Rolf, HB9SDB. See http://www.xreflector.net/

➢ XLX

XLX is the newest reflector system. It was developed by Jean-Luc Boevange, LX3JL, and Luc Engelmann, LX1IQ. XLX is an open system and it supports all of the other D-STAR protocols like DCS, XRF (Dextra), and REF (DPlus), as well as other digital modes. The XLX system can also connect reflectors together.

http://xlx000.xlxreflector.org/index.php?show=reflectors.

If you are using a D-Star radio you would connect to an XLX reflector by using a DCS or sometimes an XRF number. For example, to connect to XLX299 you use a UR code of DCS299BL. DMR users will connect to a different module with a different command.

CONNECTING TO A REFLECTOR

The connection is achieved using the 'Link to Reflector' option. Double-tap the text to the right of the 'To' icon, to show the 'TO SELECT' sub-menu. Select <Reflector> <Link to Reflector>.

At this stage the menu will show 'Direct Input' and it may list some previously selected reflectors. You can select the one you want if it is listed. Assuming that you are starting from scratch, select <Direct Input> to get to the Link to Reflector sub-menu.

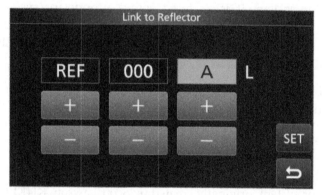

Link to Reflector sub-menu

Use the plus and minus keys to select REF, XRF, DCS, or XLS. Then the reflector number, and the module (band) letter. Then touch SET. The sub-menu will close, and you should see the correct 'Link to Reflector' text next to the 'To' icon. [In the example above, it would be REF000AL].

Before connecting to a reflector, make sure that the repeater is not linked to another repeater or a reflector. Before you unlink the repeater, put a call out and let the other repeater users know that you are planning to unlink the repeater. Someone may be waiting for a call from a linked repeater or for a Net to start on a reflector. It is good etiquette to restore the linking after you have finished using the repeater. So, check the repeater dashboard on your PC to see where the repeater is linked, or send an Information request on the radio and take note of the linking information.

To connect to the reflector, press the Microphone PTT for a second or two, there is no need to wait for the characters to scroll through.

You may get a scrolling text message and often a voice message stating that the repeater is now connected to the reflector. Once the connection has been made you should immediately touch the 'To' icon if it is not already highlighted and turn the VFO knob until the 'To' text box says, 'Use Reflector CQCQCQ.' *You can also do this the hard way by touching the 'To' text box and selecting <Reflector> <Use Reflector>.*

Listen for traffic on the reflector before calling CQ or another station.

IC-9700 DV GATEWAY

The gateway mode allows the radio to act like an Internet terminal or a hot-spot repeater. The setup is quite complicated. It is covered in detail, in section 11 of the Icom Advanced Manual.

➢ The IC-9700 Terminal Mode

The Terminal Mode lets you configure the radio to access gateway repeaters and reflectors directly over the Internet, without using a repeater or any radio transmission at all. This is not really ham radio in my opinion. More like Skype. You need an Ethernet connection and you may have to open a port on your Internet router to let the radio access the Internet.

➢ The IC-9700 Access Point Mode

The Access Point mode is like the terminal mode except that it acts as a hot-spot. The radio is connected via the Ethernet cable and the Internet to a gateway or reflector. You can use a handheld radio to talk to the receiver in the IC-9700 which passes the audio through the Internet to the distant gateway. Audio from the gateway is transmitted by the IC-9700 to your handheld radio. This is a very expensive way to implement a D-Star hot-spot, but I guess it could be useful if you wanted to make D-Star calls from around your yard.

➢ OPC-2350LU

If you can't connect the radio directly to an Ethernet port on a fiber or ADSL router connected to the internet, you can use an OPC-2350LU or similar data cable to connect the radio to a Windows PC or Android device and then use that arrangement to connect the radio to the Internet. A Windows device will need RS-MS3W software and an Android device will need RS-MS3A software. The software is a free download.

DTMF CODES

Many repeaters can use short DTMF codes to control repeater functions including linking to reflectors. You can send a DTMF code by using the DTMF screen. <MENU> <2> <DTMF>.

Touch SEND to immediately send a code using 'Direct Input,' or you can send one of the last codes sent. Touch EDIT to pre-load up to 16 saved codes. SET changes the speed that the codes are sent from the transmitter. I have changed to 200ms because my repeater seemed to miss the codes when the speed was set to 100ms. But you can try 100ms first.

➢   Common DTMF repeater codes.

Unlink repeater   #   (same as UR = _ _ _ _ _ _ _ U)

Linking information   0   (same as UR = _ _ _ _ _ _ _ I)

Reset to default link   **   (resets the gateway to its default linking)

➢   Some repeaters support DTMF gateway linking to reflectors:

(example) Link to a DCS reflector DCS001A   D1A or D101

(example) Link to a DCS reflector DCS005B   D5B or D502

(example) Link to a REF reflector REF006C   *6C

(example) Link to a REF reflector REF005A   *5A

It seems that you can only access reflectors from 001 to 009 using DTMF since the last two numerals are used to select the module letter. The letters A to Z are represented by numbers 01 to 26. To access DCS006Q you would send a DTMF code of D617.

D-STAR (DR) SCAN MODE

The DR (digital repeater) mode has a Scan function. In DR mode, press the SCAN button to scan the FROM repeater through your local channels.

The scanned channels can include FM repeaters as well as D-Star repeaters. You can also start a DR scan by pressing <MENU> <SCAN>. The Scan Mode will update the FROM box and step the radio through your local DV channels and any FM channels that are included in the DD/DV memory. You only want to scan the local channels and hotspots within the range of your radio. The ones you can trigger. All other channels and overseas groups should be excluded from the scan using the 'skip' setting. There is no point in listening for channels that are too far away to be picked up by the radio.

I found the easiest thing to do was to set 'Skip all ON' on all of the groups and then use the individual SKIP Soft Key to set my four local repeaters so they will not be skipped.

To change the Skip settings, open the DD/DV memory bank using <MENU> <2> <DD/DV MEMORY> <Repeater List>. Touch and hold a memory group title and select <SKIP ALL ON> to Skip the entire group. When you have blocked all of the groups, touch your local group and touch and hold the entry for your local D-Star repeater. On the quick menu, touch SKIP to 'un-skip' that channel. It should now be the only channel that does not have 'SKIP' on the right side of the listing. Repeat this for any other local repeaters.

➤ SCAN button in DR mode

Pressing the SCAN button in DR mode opens a sub-menu. You can select from,

- Normal – scans all non-skipped repeaters in the DV/DD Repeater List.

- Near Repeater (DV/FM) scans all non-skipped repeaters near you

- Near Repeater (DV) scans all non-skipped DV repeaters near you, or

- Near Repeater (FM) scans all non-skipped FM repeaters near you

➤ MENU SCAN in DR mode

Selecting <MENU> <SCAN> while in the DR mode opens a small scanning window. It is fairly limited. It displays (but won't let you change the setting above). You can recall a repeater frequency if one is listed in the RECAL box and you can turn the scan on or off with the SCAN Soft Key.

The SET Soft Key allows you to set scan speed, resume (stop or pause), pause timer (2-20 seconds or hold until the signal disappears), resume delay timer (wait a few seconds then resume the scan), temporary skip timer (start to scan skipped channels after a set period, or never scan them), and 'Main Dial Operation (SCAN).' I don't know what the last one does. It is not mentioned in either of the Icom manuals.

D-STAR PICTURE SHARING MODE

The picture sharing mode was introduced as a firmware update for the IC-9700 and has been included in the IC-705 as well. It allows you to send and receive photos over the D-Star simplex or repeater mode. Not wowed so far?

The neat feature is that with the free Icom ST4001 software for Android, IOS, or Windows you can take a photo with your phone, copy it to your IC-9700 via Wi-Fi and the radio's Ethernet connection, and transmit it to another station. "Here is a photo of us at the beach!" or "This is my latest toy, an icom IC-9700!"

It seems complicated initially, but it is actually quite easy.

Icom has a help file for Picture Sharing mode at,
https://www.icomjapan.com/uploads/support/manual/Picture_Tx_ENG_IM_2.pdf

PICTURE

The picture menu <MENU> <2> <Picture>, opens the 'Picture Sharing' mode. You can view pictures received over D-Star and send pictures that have been received from the ST4001 software and stored on the SD card.

While you can take the SD card out of the radio and load the images onto the card from your PC. There is a better and much cooler way! With the free Icom ST-4001 software you can download an image from your phone directly to the radio.

➤   RX

Touch the RX image to get a larger image of a received photo. You also get information about who sent it, the image size, and the quality setting that was used. Select the HISTORY Soft Key to get a full-screen image or step through previously saved images. You can also save the picture as a JPG. When you receive a transmission that includes a picture the 'RX Picture icon' will be displayed on the main display. If you happen to have the 'Picture Sharing' mode onscreen you will see the picture appear block by block. But the image will still be received even you are not in the Picture Sharing mode.

Depending on the picture size and quality, and the transmission method, it may take several overs before a complete picture is received.

➤   TX

The TX image is shown as a group of blocks. Each block is sent as a data burst. If you select a high resolution, there are more blocks, so the image takes longer to send. This does not matter for simplex contacts, but it could annoy D-Star repeater or reflector users. The top left square should be red indicating that it is the first block to be transmitted. The red block moves across and down the picture as the image is being sent. To select a different image, touch the TX image, then touch it again or TX

SET, then touch the TX Picture line and use the 'up' and 'down' arrows to select a different picture. Finally, touch the SET Soft Key to select it. Then touch Back ↺, Back ↺.

TX SET lets you change the picture size. The default size is 320x240. You can also send at three different quality levels. I suspect this represents three levels of JPG data compression.

RECEIVER. You can change the data that would normally go in the TO box (UR data field). CQCQCQ will send the picture to anyone who has the capability to receive it. Or you can enter a destination callsign. Or you can select one of your buddies from the rather misnamed 'Your Call Sign' group.

HISTORY shows you images you have transmitted before and allows you to re-transmit them.

BLOCK lets you change the start position in case you want to leave out the top part of the image.

1ST. If the red block is not top left and you want to start again. Touch and hold the '1ST' Soft Key.

ⓘ Indicates the IP address and network name (if set). This is only relevant if you are using the Ethernet connection to upload photos into the radio.

➤ PICT TX

If you select PICT TX, (while in D-Star mode), an icon will appear at the top of the display indicating that a picture will be transmitted with your next D-Star transmission. The picture will be sent along with your voice transmission and will appear on the Picture RX screen at the destination radio. When you stop transmitting, the picture will pause. On your next over, the picture transmission will continue from where it stopped on the previous transmission.

So, it may take several overs before a full image (particularly a large or high-quality image) is completed.

➤ TX ALL

The TX ALL mode sends the whole picture as a high-speed data transmission. It is much faster than the PICT TX mode, but you cannot talk at the same time. Announce that you are going to send a picture, then send it using TX ALL.

GPS location

GPS location information is used for three things. It can be transmitted with your D-Star transmissions so that the station you are calling can see; your location, Maidenhead grid reference, distance, speed, the direction you are heading, and what direction you are from them. It is also used for the 'Near Repeater' function which lets you choose DV or FM repeaters that are near your location. And it is used to determine the distance and heading towards stations that you hear via the repeater or gateway.

If you want to run D-Star but you don't plan to operate mobile or portable, there is little point in connecting an externals GPS. Just enter your location manually.

A GPS referenced clock can be used as a very accurate frequency reference that can be applied to the 10 MHz reference signal input for the purpose of 'governing' or calibrating the radios TXCO. See the section on connecting a GPS reference or atomic clock on page 207.

CONNECTING A GPS RECEIVER

If you plan on operating in D-Star mode while mobile or portable, you can connect the radio to an external GPS unit. It has to be NMEA compatible and it has to output at RS-232 data levels. The expected data rate is selectable, but the default is 9600 bps. The GPS receiver is connected to the 2.5mm 'Remote' jack on the rear panel.

I initially tried a tiny GPS unit that I had been using with an Arduino, but the I2C data level is too low. So, I ended up buying a cheap GPS module from the Alexnld.com Internet site. The module is a Beitian BS-280. It outputs NMEA-0183 data at 9600 bauds at RS-232 levels and it cost $18 USD including postage to New Zealand. The unit requires a 5V DC supply. I powered mine from a USB hub. The connection to the radio is a 2.5 mm stereo mini phono connector. RxD (green) on the GPS is connected to the ring, TxD (white) is connected to the tip and the earth is the black wire.

Set the GPS to 'External GPS' and the baud rate to suit the GPS unit. When you return to the main radio screen, you should see a little satellite icon at the top of the screen. It will flash if it is receiving data but has not yet seen enough satellite data. It stops flashing after data has been received from several satellites. This may take several minutes. If you don't see the satellite icon, the chances are that you have the RxD and TxD connection reversed, or the level from the unit is not sufficient.

TIP: The GPS receiver will also keep the clock display accurate. Select <MENU> <SET> <Time Set> <Date/Time> <GPS Time Correct> <Auto>.

➤ GPS Information

<QUICK> <GPS Information> is a quick way of displaying the GPS information screen. It is only valid if you have a GPS receiver connected to the radio. It is the same screen as you get through the main menu GPS sub-menu but it easier to access from the QUICK button.

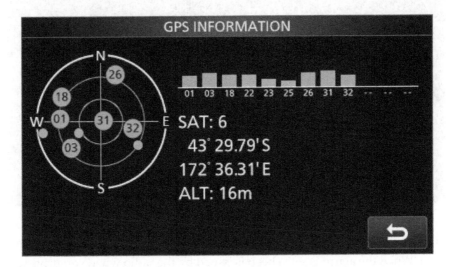

The GPS Information screen shows the satellite number and bearing of satellites being received by the connected GPS receiver. The image is a 'Radar Plot.' The center is your location. Satellites that are high overhead are shown close to the center. The outer ring is the horizon. The white dots are GPS satellites that are being received but the data is invalid, usually because of extreme range. If no GPS receiver is connected, there is no display.

➤ GPS Position

<QUICK> <GPS Position> is a quick way of displaying the four GPS position screens. It is the same screen as you get through the main menu GPS sub-menu but easier to access from the QUICK button.

The GPS Position screen shows your location. Latitude, longitude, Maidenhead Grid, altitude (can be inaccurate), your speed, course, and the current time.

The second screen shows the GPS data from a received D-Star signal.

The third screen shows the GPS data from a location stored in the GPS memory bank. Touch and hold the screen to load a location from the GPS memory.

The fourth screen can show the GPS Alarm distance and bearing to a nominated location. When the alarm is set, a beep is issued by the radio when you drive within 1 km of the nominated site or group of GPS locations. At 500 meters three beeps are issued. This is a function known as 'geo-fencing,' (advanced manual page 9-26). It can also be used to alert you when a 'target' station in your list comes within 1 km or 500 meters of your location or within the selected area. The station has to transmit within the area for the alarm to work.

ENTERING YOUR LOCATION MANUALLY

The first thing you need to know is the latitude and longitude of your location. You can get this from a phone App like 'Maidenhead,' Google Earth or Google Maps (click on the map and look at the URL line).

TIP: Since firmware update 1.20 you can also enter latitude and longitude as degrees with decimal points.

To enter your location manually select <MENU> <2> <GPS> <GPS Set> <Manual Position>. Touch and hold anywhere on the screen, then select 'Edit.'

TIP: If you have a GPS receiver connected and External GPS selected, you can load the manual data from that, or you can load it from a saved GPS memory.

Touch 'Latitude' and enter the latitude. If you live in the southern hemisphere remember to change N to S using the N/S Soft Key. Touch ENT to save and exit.

Touch 'Longitude' and enter the longitude. Remember to set the East-West setting using the E/W Soft Key. Touch ENT to save and exit.

Touch 'Altitude' and set your altitude above sea level if you know it. Or just set it to zero. Touch ENT to save and exit.

➢ Checking the GPS data for your station

You can check the data using Tab 1 of the GPS Position screen. <QUICK> <GPS Position> 1 of 4. You will see that the radio has calculated your Maidenhead Grid locator from the entered latitude and longitude data. If that is radically different to what you expect, you probably have the East/West or North/South setting wrong.

GPS MENU SETTINGS

See the <MENU> <GPS> chapter, page 124.

➢ GPS Set

GPS Set is used to select; OFF if you are not using D-Star at all, 'Manual' if you don't have a GPS receiver attached, or 'External GPS' if you do have a GPS receiver attached.

➢ GPS TX Mode

GPS TX Mode set whether you will transmit your position with your D-Star transmissions. And if so, whether the data is sent in NMEA or D-PRS format.

➢ GPS Memory

The GPS Memory screen provides access to the GPS memory bank. The bank can store 300 GPS locations divided into 26 groups.

➢ Alarm

GPS Alarm is used to set the GPS alarm geofencing parameters.

➢ GPS Auto TX

GPS Auto TX can be set to send your location data over D-Star at regular intervals ranging from every 5 seconds to every 30 minutes.

Satellite Mode

Using the IC-9700 as a ground station transceiver is one of the most important features of the radio. Many hams will buy the radio specifically for Satellite operation. The IC-9700 offers full-duplex cross-band satellite mode operation with forward and reverse tracking of the VFO frequencies.

The tracking mode locks the transmit band VFO and the receive band VFO together so that both change as you tune the VFO knob. Tracking works when tuning either the transmit or the receive VFO. With 'reverse tracking' turned on, the receiver frequency will decrease as the transmit VFO is tuned higher and vice versa. This mode is used when operating through 'reverse tracking' SSB transponders. With 'normal' tracking enabled, the receiver frequency will increase as the transmit VFO is tuned higher and decrease as the transmit VFO is tuned lower. The normal tracking mode is used when operating through FM, Digital, or 'normal tracking' SSB, transponders.

The built-in satellite mode has an innovative way that the transmit or receive VFOs can be quickly unlocked from tracking each other, adjusted for Doppler shift, and then locked back into tracking mode. This makes manual tracking of FM transponder satellites easier to accomplish. Being able to see the downlink signal from the satellite on the panadapter display makes it very easy to tune your receiver to the signal and follow the Doppler shift as it changes the frequency of the downlink signal. For SSB transponders, most users will probably prefer to use satellite tracking programs which control the radio through the CI-V interface to automate Doppler correction and provide transponder tracking.

I have seen suggestions online that satellite operation with PC control should be done in the 'normal' rather than the 'satellite mode,' but I believe that you should use the satellite mode, especially for SSB transponders.

I successfully configured the radio to work with SatPC32 and others have reported success with MacDopplerPRO. One feature not found on other 'satellite capable' transceivers is that you can listen to the uplink frequency and the downlink frequency at the same time.

SATELLITE MODE

Turn on the satellite tracking mode by pressing <MENU> and then selecting the <SATELLITE> icon at the top right of the menu screen. This activates the last used satellite frequency pair, with the downlink frequency on the upper VFO display and the uplink frequency on the lower VFO.

Three Soft Keys are displayed to the right of the screen. They enable or disable forward or reverse VFO tracking.

The tracking function in the radio is more about tuning across a transponder bandwidth to find signals than correcting for Doppler shift. It would be nice if there was a 'Doppler' Soft Key button on the screen that made the 70 cm VFO increment at a rate three times faster than the 2m VFO, as that is close to the relative Doppler shift. It wouldn't completely eliminate the need for manual Doppler correction on SSB, but it would certainly help. I believe that such a control would be completely adequate for Doppler correction when using FM transponders.

The published transponder frequencies for satellites are the frequencies sent from and received at the satellite. They must be adjusted to compensate for Doppler shift. The frequency received at your location will be higher than the published downlink frequency at the start of the satellite pass and lower than the published downlink frequency at the end of the satellite pass. To reach the satellite on the correct frequency you must transmit low at the start of the pass and high at the end.

- VU mode means a 2 m band uplink and 70 cm downlink
- UV mode means a 70 cm uplink and a 2m downlink.

RECEIVER AND TRANSMITTER VFO POSITIONS

In satellite mode, the downlink (receiver) VFO is on the top and the uplink (transmitter) VFO is on the bottom. The is the reverse of the normal radio mode where the VFO that is able to transmit is always in the top display position. I have no idea why Icom decided to arrange the VFOs differently for the satellite mode. Perhaps, having the downlink on VFO-A makes the setup more complaint with external satellite tracking programs and previous Icom models.

SATELLITE MEMORY BANK

While in the satellite mode, the standard memory <MENU> <MEMORY> holds the 99 satellite memory slots. These are accessed and stored in the usual way. Use the V/M button to select memory mode. If M-CH is selected, the MULTI knob switches through the channels. The channel name is displayed next to the FIL icon on the lower VFO.

Handy TIP: I found a list of amateur radio satellite frequencies on the https://groups.io/g/ic-9700/topics website and downloaded them to the radio using the Icom CS-9700 software.

The frequencies stored in the memory slots are usually the frequencies received and sent from the satellite. In other words, you need to adjust them for Doppler shift as the pass progresses.

MANUAL TRACKING – FM SATELLITE TRANSPONDERS

Frequencies are not as critical when you are using FM satellite transponders. I believe the easiest method is to save a series of five memory channels and switch through them during the satellite's pass. As the received signal starts to sound noisier, switch to the next channel. If the satellite pass has a low peak elevation you might only use the three middle channels.

Table 1: Channel table for a typical VU mode FM satellite

Channel	Transmit	Uplink offset	Receive	Downlink offset	Time approx
1	144.348 MHz	- 2 kHz	437.208 MHz	+ 8 kHz	AOS
2	144.349 MHz	- 1 kHz	437.204 MHz	+ 4 kHz	+2 mins
3	144.350 MHz	0	437.200 MHz	0	+4 mins
4	144.351 MHz	+ 1 kHz	437.196 MHz	- 4 kHz	+6 mins
5	144.352 MHz	+ 2 kHz	437.192 MHz	- 8 kHz	+8 mins

Table 2: Channel table for a typical UV mode FM satellite

Channel	Transmit	Uplink offset	Receive	Downlink offset	Time approx
1	437.192 MHz	- 8 kHz	144.352 MHz	+ 2 kHz	AOS
2	437.196 MHz	- 4 kHz	144.351 MHz	+ 1 kHz	+2 mins
3	437.200 MHz	0	144.350 MHz	0	+4 mins
4	437.204 MHz	+ 4 kHz	144.349 MHz	- 1 kHz	+6 mins
5	437.208 MHz	+ 8 kHz	144.348 MHz	- 2 kHz	+8 mins

Doppler never sleeps. When using the 70 cm or 23 cm bands you always need to take Doppler shift into account. Doppler can cause a frequency shift of as much as ±10 kHz on the 70cm band. There is no point listening for a satellite if you are tuned several kilohertz off frequency. However, because the maximum Doppler shift on the 2 m band is only about 3 kHz you can usually get away with leaving it set to the satellites published frequency throughout the pass. But since this radio has the capability to store channels with less than 5 kHz channel spacing, we might as well use it.

During the pass, you can unlink the VFOs and tune the downlink (receive) VFO a little, but don't forget to change channels every few minutes, or the uplink signal may drop out of the satellite's receiver bandwidth. Especially if it is a UV mode satellite.

MANUAL TRACKING - USING THE VFOS

The technique for manual tracking starts with setting the uplink and downlink frequencies to the published (or known) frequency pair. You will adjust the frequencies during the satellite pass using a combination of manual VFO control and tracked VFO control. The downlink frequency should be entered into the top VFO and the uplink frequency is entered on the lower (transmitting) VFO. It is a good idea to save the frequency pair in one of the 99 satellite memories so that they are there next time you want to work the same satellite and in case you get way off frequency while working the satellite.

For a VU mode satellite where the downlink signal is on the 70 cm band, unlink the receiver VFO and tune the receiver frequency up by 8 or 9 kHz so that you ready for the start of the satellite pass. Leave the 2m band transmit frequency alone during the pass. Tune your receiver until you hear a station using the satellite, or if nobody is using the satellite until you can hear your transmitted signal coming back from the satellite. Now you can make a CQ call. Keep slowly tuning the downlink (receiver) frequency lower during the pass. If there is traffic, tune for the best signal, if not the receive frequency should pass through the published (saved to memory) frequency when the satellite reaches peak elevation and the received frequency will continue to drop until the satellite sets below your local horizon.

Tuning the downlink signal is easier than setting the uplink frequency because you can hear the signal and see the signal trace on the waterfall display. This makes VU mode satellites easier to work with manual tuning than UV mode satellites because you can leave the 2 m band uplink frequency alone and just adjust the received downlink signal for Doppler correction.

For a UV mode satellite where the uplink signal is on the 70 cm band, you will leave the downlink receive signal alone until you hear a signal from the satellite. After that, you can adjust it by ear or by observing the received signal on the waterfall trace. Start by tuning the receiver 2 – 3 kHz high so that you are ready for the start of the satellite pass. The transmit frequency will have to be adjusted from low to high as the pass progresses. Start with the frequency about 8 to 9 kHz low and if you are getting into the satellite you should see your signal on the downlink frequency as a line on the waterfall. Since the setup is full-duplex you should also be able to hear your transmission on the downlink receiver.

If you are using SSB mode through a linear transponder on the satellite, it is MUCH easier to use a satellite tracking program to control the Doppler correction.

At the start of the satellite pass, set the downlink frequency high and the uplink frequency low to account for the expected amount of Doppler shift. Turn tracking on and tune the downlink receiver until you hear a station calling CQ or a quiet area on the transponder where you can make a call. This is the only time that the VFOs should be linked. Generally, users of SSB will spread outwards from the transponder center frequency as more users come onto the transponder. CW and other modes are usually on the lower end of the transponder downlink band. Once you have selected a downlink frequency, turn the tracking off. Talk on the uplink and tune the uplink frequency a little until you can hear yourself on the downlink. Keep overs very short and adjust the downlink frequently.

Even if you are not using computer-controlled Doppler compensation, you can use a satellite tracking program to tell you the bearing and elevation of the satellite and if the program features Doppler correction you can set the VFOs to the frequencies suggested by the tracking program.

Tracking (when used) will be 'normal' (NOR) tracking for an FM transponder or a 'forward' or 'normal tracking' linear transponder. It could be 'reverse' for an SSB 'inverting' or 'reverse tracking' linear transponder.

Overall while manual tracking is OK for FM satellites, if you are using SSB mode through a linear transponder on the satellite, it is far easier to use a satellite tracking program to control the Doppler correction. And if you are set up for computer tracking, you might as well use it for FM and digital mode satellite operation as well.

SETTING UP THE RADIO FOR PC SATELLITE OPERATION

The setup for the radio is the same as the setup for digital modes. See page 23 for details on how to configure the USB connection between the radio and the PC. The audio connections probably won't be required.

SATPC32

I have tried out quite a few satellite tracking programs and found that SatPC32 works best, for me.

SatPC32 is available as a download from http://www.dk1tb.de/indexeng.htm and also from the AMSAT-NA, AMSAT-UK, and AMSAT-DL websites. It was written by Erich Eichmann, DK1TB. He has very kindly donated all the proceeds from the software to AMSAT, to support amateur satellite development. The software is available in English or German language versions. If you use the dk1tb link, you can download a trial version for free and register it later via any of the above mentioned AMSAT websites. The program has a few quirks, particularly when it comes to storing the frequencies that are used by the satellites, and configuring antenna tracking units, but it provides excellent control of the radio, accurate tracking, and accurate Doppler shift compensation.

SATPC32 SETTINGS

SatPC32 does not directly support the IC-9700, but the program works very well with the following settings.

1. On the [Setup] tab select [Radio Setup]

2. Change the 'Radio 1' radio button to Icom

3. Set the Com port to the one used by your radio. Click on the Com Port number and enter the new value into the text box to the right.

4. Leave the CAT delay at 40. If you do want to change it, click on the CAT delay number and enter the new value into the text box to the right

5. Check the 'Automatic RX/TX change' and 'Satellite Mode' checkboxes

6. Use the dropdown list under [Model] to select IC-9100

7. Click the 'Model' dropdown box and select 'Baudrate.' In the lower dropdown select the baud rate the radio is using. If the radio is set to 'Auto,' select 19200

8. Click the 'Model' dropdown box and select 'Addresses.' In the lower dropdown select the radio address. The default (recommended) address is $A2 $A2. This should change the text box below the dropdown to $A2 $A2. If the dropdown does not have the wanted address, you can type it into the lower text box, next to where it says Kenwood.

9. Click the 'Addresses OK' button

10. Click the 'Store' button

You will be able to see when the program is doing Doppler correction on the radio. Every now and then the frequency of the upper VFO will update and the display will flick down to the other the VFO and update that.

I like to set the MODE settings in SatPC32 for Doppler correction of both the 'Uplink and Downlink.' You have to select an SSB or CW satellite transponder before you can change this setting. The RX/TX addresses should be $A2 (unless you have changed the radio's address). If you are using SSB transponders, set 'Display in Frequency Window' to 'Frequency at Satellite.' This will display the frequency at the satellite below the 'Observer' frequencies at your location. It will let you see where you are on the linear transponder. If you select 'Doppler Shift' the amount of Doppler shift correction is displayed below the 'Observer' frequencies, instead of the frequency at the satellite.

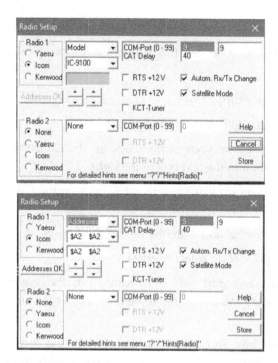

SatPC32 radio settings

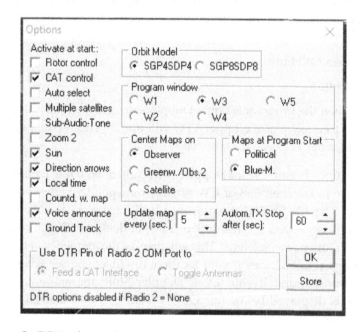

SatPC32 other settings

SATPC32 OPERATION

Most of the notes in this section will apply to any CI-V satellite tracking program.

➢ Linear transponders

SatPC32 should be in V+ mode for linear transponders. If you have set the NOR/REV tracking on the radio, the V+ option will allow you to tune across the transponder. For FM, digital mode and any other single frequency

```
File   Tracking   Satellites   CAT
B: AO-07
R- C+ A- U+ T0 L AL CW-
M- Z1 G- S+ D+ W3 BM 2D
```

satellite transponder the V- setting should be selected. C+ indicates that CI-V control is active. T0, T1 or T2 indicates whether a CTCSS tone is turned on.

➢ VFO update rate

When the SatPC32 tracking program updates the radio with corrected frequencies, it loads the downlink VFO and then the uplink VFO. This causes the radio focus to switch to each VFO, in turn, creating a rather distracting flash of light. I found that it is best to decrease the rate that the program updates the radio so that these updates are less intrusive. You do this by telling SatPC32 how many Hertz the frequency can change before it issues an update to the radio. Use the CAT tab and set the 'Interval' for 'SSB/CW' between 10 Hz and 50 Hz and set 'FM/PKT' to 500 Hz. The radio will be updated more often in the middle of the satellite pass where the rate of change is higher. At peak elevation, there is no Doppler shift and your receiver will be tuned to the same frequency as the one the satellite is transmitting. The amount of Doppler shift is at a maximum at the beginning of the satellite pass (AOS) [acquisition of satellite] and end of the pass (LOS) [loss of satellite]. It is dependent on the altitude of the satellite's orbit because the satellite's altitude governs the speed that it travels around the Earth. For a satellite in a typical low Earth orbit, the maximum Doppler shift will be less than 3 kHz on the 2 m band, 9 kHz on the 70 cm band, and 29 kHz on the 23 cm band.

If you believe that the Earth is flat, amateur radio satellite operation is "not for you."

➢ Kepler data

Kepler or TLE (two-line element) data is the set of numbers that the satellite tracking software needs in order for it to correctly calculate the time and bearing that the satellite will rise above your local horizon, the path it will follow to the peak elevation, and the path and bearing where the satellite will set below your local horizon again. The same information is used to calculate the relieve speed between the satellite and your location so that the instantaneous amount of Doppler shift correction can be applied to your radio's VFOs. SatPC32 can update the TLE data for all of the amateur radio satellites automatically, by downloading the 'amateur.txt' data file from the Internet.

➢ Satellite transponder frequencies.

Unfortunately loading the transponder frequencies into SatPC32 is not as easy. You must manually update a text file called 'Doppler.sqf.' If the satellite has several transponders, or you want to enter beacon frequencies, you add a line for each one.

The format is downlink frequency, uplink frequency, downlink mode, uplink mode, normal or reverse tracking, downlink offset, uplink offset, and note. For example; **FO-29,435910,145850,FM,FM,NOR,0,0,FM Repeater.** The note is not displayed other than on the "Data Line" in the CAT tuning selection sub-menu in SATPC32 and in the SQF file.

If you are only interested in the downlink. For example when you are saving a beacon frequency. Just enter the comma delimiter or a zero in place of the uplink frequency. For example: XW-2C,145770,0,FM,,,0,Digital TLM.

➢ Tones

SatPC32 can store the CTCSS tones used to access some FM satellites. The tone can also be stored in the radio's satellite memory slots with the rest of the channel information. Tone operation is the same as for accessing ordinary repeaters, except most people only use tones on the uplink and not on the downlink. Use 'TONE' not 'TSQL.'

The SO-50 satellite uses two different tones. A tone of 74.4 Hz is sent for a couple of seconds at the start of the pass to turn on a 10-minute timer on the satellite. After that, you must send a tone of 67 Hz with all of your transmissions to open the receiver squelch at the satellite. The easiest way to do this is to store an additional memory position with the frequencies used at the start of the pass, and the 74.4 Hz tone for the uplink.

TIP: Many operators leave the receiver squelch open (noisy) during satellite reception. Make sure that you are listening to the noise from the downlink (upper) VFO and not the uplink (lower) VFO. Turn down the volume or turn on the squelch on the lower VFO.

MORE INFORMATION ABOUT SATELLITE OPERATION

Satellite operation is a big topic. Much too big to be covered in a single chapter of this book. If you are interested in getting started with this fascinating aspect of the amateur radio hobby, please consider buying my book about amateur radio satellite operation, **'Amsats and Hamsats - Amateur Radio and other Small Satellites.'** The book has step by step guides to satellite operation as well as explaining how satellites are launched and why they stay in orbit. There is also a condensed history of amateur radio satellites and key events such as the first satellite, the first US satellite, and manned space missions carrying amateur radio into space.

Useful Tips

This chapter includes some techniques which may be useful, or at least interesting. I hate the default setting for the spectrum scope, so the first item in the chapter is a method of setting up the scope colors in a way that makes the display look like the panadapter display of most SDR receivers. The second item is a small paragraph about the screen-saver. I included it because I didn't even know there was a screen saver. The third item is a discussion on the mode switching. I can see why Icom configured it the way that they have. But I find it confusing. Considering that it is just a window on a touch screen and a bunch of Soft Keys I don't know why they didn't just add individual Soft Keys buttons for the SSB-D, AM-D, and FM-D data modes. The last item is about turning off the AGC. I don't recommend ever doing that, but I was intrigued to find that it is possible.

SETTING THE SPECTRUM VIEW TO A LINE VIEW

Unlike the audio spectrum display, the main spectrum scope display can't be set to a line rather than a filled-in trace. I prefer to see a line, similar to the display on a spectrum analyzer or almost every other SDR panadapter. Luckily careful adjustment of the screen colors allows you to fake it so that the filled-in trace looks like a line display.

The spectrum can be set to show a filled spectrum or a filled spectrum with a line of a different color on the top. There is also a max hold display and you can set the color of that as well. Personally, I hate the max hold display and always turn it off.

To display a spectrum line instead of a filled spectrum, try the following settings.

1. These settings set the peak of the spectrum waveform to a white line and the filled in part of the display to a blue that is slightly darker than the waterfall giving the appearance of an unfilled spectrum trace.

2. Hold the M.SCOPE button for one second, until the panadapter and associated Soft Keys are displayed.

3. Touch and hold the EXPD/SET Soft Key down for one second to display the settings.

4. Scroll to display 'Waveform Type' and set it to 'Fill+Line' then touch the Return icon ↩.

5. Touch 'Waveform Color (Current).' Set the slider controls for 30 red, 30 green, and 191 blue. Then touch the Return icon ↩. You can set the colors back to default settings by touch and holding 'Waveform Color (Current).'

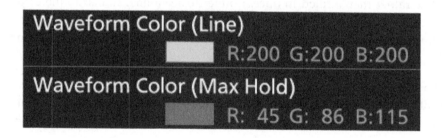

Waveform Type
Fill+Line
Waveform Color (Current)
R: 30 G: 30 B:191

6. Touch 'Waveform Color (Line)' and set the controls for white. I have mine set a little less bright at, 200 red, 200 green, and 200 blue. Then touch the Return icon ↩. You can set the colors back to default settings by touch and holding 'Waveform Color (Line).'

7. I don't use the Max Hold indication, but if you do, you can probably leave the 'Waveform Color (Max Hold)' colors at the default, 45 red, 86 green, and 115 blue.

Waveform Color (Line)
R:200 G:200 B:200
Waveform Color (Max Hold)
R: 45 G: 86 B:115

SCREEN SAVER MODE

The radio has a screen saver function to preserve your display just like your PC does. The screen goes blank and the green LED on the Power button flashes to indicate that the radio has gone into screen saver mode. Touch the screen or press any button to restore normal operation.

You can turn the screen saver function off (not recommended) or adjust the delay before the display goes to sleep using <MENU> <SET> <Display> <Screen Saver> <OFF, 15 min, 30 min, or 60 min>.

Note that received audio and the output to digital mode software via the USB cable will continue even though the screen is black.

RETURN TO SSB/FM/AM FROM A DATA MODE.

You would think that changing back to a voice mode from a data mode would be straightforward like selecting AM, FM, or SSB. But it does not work that way. The Data modes are considered to be sub-modes of the three voice modes. USB-D is a sub-mode of USB, FM-D is a sub-mode of FM, and AM-D is a sub-mode of AM.

- To return to USB from USB-D, don't select SSB mode. Touch DATA again.

- To return to LSB from LSB-D, don't select SSB mode. Touch DATA again.

- To return to FM from FM-D, don't select FM mode. Touch DATA again.

- To return to AM from AM-D, don't select AM mode. Touch DATA again.

Why didn't Icom just add four additional Soft Keys for the four data modes?

TURNING THE AGC OFF - AGC TIME CONSTANT

The AGC Soft Key on the FUNCTION menu cycles through Fast, Mid, and Slow.

The AGC time constants are different for each mode. In the FM or DV modes, the AGC is always set to Fast and you can't adjust the AGC time constant.

Touch and holding the AGC Soft Key brings up a screen that allows you to change the AGC settings. The only way to exit this screen is to press the EXIT button.

- To adjust the time constant, select, Fast, Med or Slow for the mode you want to change.

- Turn the main VFO knob to change the time constant. You can even set the AGC to Off. But other than for lab testing I don't know why you would want to do that.

- You can change the operating mode without closing the AGC adjustment screen and adjust the AGC settings for all modes in turn.

- The only way to exit the AGC adjustment screen is to press the EXIT button.

Troubleshooting

The items covered in the troubleshooting chapter are not faults. However, they are conditions that might worry you if you encounter them while operating the radio. This chapter may help you if something unexpected happens.

BLACK SCREEN AND A GREEN LED FLASHING ON THE POWER BUTTON

Don't panic! The radio has gone into screen saver mode. The radio has a screen saver function to preserve your display just like your PC does. Touch the screen or press any button to restore normal operation.

The green LED on the POWER button flashes to indicate that the radio has gone into screen saver mode.

You can turn the screensaver off, but I don't recommend doing so. You can change the time before the screen saver turns the screen off using <MENU> <SET> <Display> <Screen Saver> <OFF, 15 min, 30 min, or 60 min>.

WATERFALL TOO DARK, ONLY LARGE SPECTRUM PEAKS ARE SHOWING ON THE SPECTRUM DISPLAY.

Sadly, the panadapter REF level is the same for all bands, so if you change bands you are very likely to have to adjust the reference level.

If there are no Soft Keys at the bottom of the display press and hold the M.SCOPE button until they appear. It is easiest if you adjust the reference with the largest panadapter. Turn the Sub Receiver off and touch EXPD/SET to get the big panadapter with the small VFO display. If there is no REF Soft Key at the bottom of the display, touch <1> to change the menu.

Touch the REF Soft Key and turn the main VFO knob clockwise to adjust the level until the spectrum noise floor is just visible at the above the waterfall. The waterfall brightness should now be correct. The reference level is indicated in a popup window. Touch REF again to exit the setup screen.

WATERFALL TOO LIGHT.

Sadly, the panadapter REF level is the same for all bands, so if you change bands you are very likely to have to adjust the reference level.

If there are no Soft Keys at the bottom of the display press and hold the M.SCOPE button until they appear. It is easiest if you adjust the reference with the largest panadapter. Turn the Sub Receiver off and touch EXPD/SET to get the big panadapter with the small VFO display. If there is no REF Soft Key at the bottom of the display, touch <1> to change the menu.

Touch the REF Soft Key and turn the main VFO knob anti-clockwise to adjust the level until the spectrum noise floor is just visible at the above the waterfall. The waterfall brightness should now be correct. The reference level is indicated in a popup window. Touch REF again to exit the setup screen.

SMALL PANADAPTER HAS NO SOFT KEY CONTROLS.

The smallest panadapter has no associated Soft Key icons. Press and hold the M.SCOPE button until the medium size panadapter appears. Unfortunately, doing this will close the Keyer, Voice, or Decode window if it is open.

DATA MODE SELECTION CONFUSION

Data mode selection is initially a bit confusing. With the other modes, you just touch the mode you want. For example, to change from SSB to RTTY you touch RTTY. To change back to SSB you touch SSB. But this does not work if you touch DATA.

The Data modes are considered sub-modes of SSB, AM, or FM and the switching arrangement is different.

1. For a start, you can only select the DATA mode if the current mode is SSB, FM, or AM. If the current mode is PSK, RTTY, or CW, the DATA mode is not displayed. To get to the SSB-Data mode select USB or LSB first. To get to the FM-Data mode select FM first. To get to the AM-Data mode select AM first. Then touch DATA.

2. To get out of a DATA mode touch DATA, **not** SSB, AM, or FM.

TUNING RATE SPEEDS UP

The tuning rate may speed up as you turn the main VFO knob. This is normal. It is linked to the speed that you turn the knob. The feature is designed to get you to the other end of the band quickly when required. If you don't like it, you can turn it off or make it slower using <MENU> <SET> <Function> <MAIN DIAL Auto TS>.

The tuning speed also changes when you are in 'fast mode' which has a 1 kHz step size. This is indicated by a small white triangle above the kHz number of the frequency display. Touch the three-number 'kHz' group on the VFO display to toggle between fast and normal mode. Touch and hold the three-number 'kHz' group to set the 'fast' tuning step for the current mode.

Touch and hold the Hz digits of the VFO display to change between 10 Hz tuning steps to 1 Hz tuning steps. In data modes, RTTY, and CW you use <FUNCTION> <¼> to set the tuning to ¼ rate to slow down the VFO rate even more.

VFO KNOB WON'T CHANGE THE FREQUENCY

This caught me out the first time I experienced it. There are two likely possibilities. You may have locked the VFO with the Speech/Lock button, but it is much more likely the radio is set to one of the two call channels C1 or C2 using the CALL DR button. Check that the text to the right of the frequency display says VFO or MEMO. Press the CALL DR button to clear the selection. This problem can occur when you select the DR mode if you don't press and hold the button long enough to engage the DR mode.

SPECTRUM ON FIX DISPLAY MODE BUT NO MARKER

This can happen if you change band EDGEs on the panadapter, or use a memory channel, or change bands. It can also happen if you just tune off the end of the panadapter. The radio is able to operate on frequencies that are not currently being displayed on the panadapter spectrum scope display.

Two small green arrows ◄◄ at the left side of the panadapter indicate that the VFO is tuned to a frequency below the currently displayed spectrum. Two small green arrows ►► at the right side of the panadapter spectrum display indicate that the VFO is tuned to a frequency above the currently displayed spectrum.

TRANSCEIVER STUCK IN TRANSMIT MODE

➢ **Starting PC Digital Mode or CS-9700 software causes the transceiver to go into transmit mode.**

This is not a fault, but it can be rather disconcerting. In SSB mode it is unlikely to damage anything as there will be very little power transmitted. But in CW AM or FM mode, it will cause power to be transmitted.

If this happens with the CS-9700 or RT Systems software, temporarily set 'USB Send' to OFF. Then set it back to USB(A) RTS when you have finished.

The problem is due to the RTS / DTS settings on the 'A' COM port being inappropriately controlled by the PC software. Normally the transceiver is set so that 'USB SEND' is set to 'USB (A) RTS. The RTS signal is used for PTT, it turns the transmitter on. The USB (A) DTS signal is used to send CW or RTTY data. If the digital mode software holds the RTS and/or DTS lines 'active' or 'always on' it will make the transceiver switch to transmit.

If this happens, check the COM port settings in the digital mode software and make sure that the RTS and DTR lines are set the same as the transceiver, or to 'always off.' The condition can be proved out of the radio by removing the USB cable. If you can't fix the settings, it is often possible to set the digital mode software to use a CAT command for PTT rather than using the RTS (or DTR) line. But I prefer the RTS and DTR method.

➢ Other reasons the transmitter may be stuck on.

If it is not the Com Port RTS / DTR settings, check that the front panel TRANSMIT button has not been pressed. The radio could also be held on transmit by the Morse key or the microphone PTT. Check that the Mic PTT button is not stuck on transmit and temporarily remove any cables from the ACC jack and the KEY jack. The PTT pin on the ACC jack also has the capability to turn on the transmitter. Basically, unplug all external cables except the antenna and the DC supply until the problem is resolved.

SPLIT NOT WORKING

The Split function will not work in the memory channel mode. Only in the VFO mode. Press V/M to change between modes.

SPLIT NOT SAVED IN MEMORY

The standard memory bank will not store the split or VFO-B stetting. If this is a problem, you can store a channel with a duplex (DUP+ or DUP-) setting equal to the split that you want. For example, +1 kHz or +5 kHz. See the next section.

NON-STANDARD OR REVERSE DUPLEX REPEATER SETTINGS

You can set the duplex setting to anything you want. It does not have to be the standard shift for the band. This can be used for odd-ball repeater shifts or to store split settings. Set the radio for the receive frequency that you want.

- Press <FUNCTION> <2> <hold DUP>
- Touch <DUP> to set a plus or minus shift
- Set the duplex split in MHz.
- Touch ENT to save and exit
- Save the channel to a memory slot in the usual way.

RECEIVING TWO FREQUENCIES ON THE SAME BAND

No, the Icom IC-9700 cannot receive two frequencies on the same band at the same time. If you are operating with the radio in Split mode, you can quickly check the transmit frequency by holding down the XFC button.

CAN'T SELECT A DV REPEATER ON A DIFFERENT BAND

The IC-9700 has two receivers but they can't both be set to receive on the same band at the same time. If you have both receivers running, you will not be able to select a repeater channel that operates on the band being used by the other receiver. If you turn off the second receiver you can select any repeater. For example, say that I am using a DV repeater on the 2 m band and I am also monitoring the local 70 cm FM repeater. I will not be able to select a 70 cm DV repeater on the Main receiver unless I turn off the Sub receiver first, or I set it to the 23 cm band. However, you can put both receivers into DR mode and listen to DV repeaters on two bands at the same time.

CW MODE SIDETONE BUT NO TRANSMIT POWER

This one has caught me out several times. The CW mode won't automatically transmit unless full or semi break-in has been selected. In CW mode press the VOX/BK-IN button and choose either BKIN (semi break-in) or F-BKIN (full break-in) mode and the radio will transmit.

The following settings affect the sending of keying macro messages as well.

- With BK-IN OFF you can practice CW by listening to the side-tone without transmitting. Or you can key the PTT line by pressing TRANSMIT, pressing the PTT button on the microphone, via a CI-V command, or by switching the SEND line on the ACC jack.

- Full break-in mode F-BKIN will key the transmitter while the CW is being sent and will return to receive as soon as the key is released. This allows for reception of a signal between CW characters.

- Semi break-in mode BKIN will key the transmitter while the CW is being sent and will return to receive after a delay when the key is released. In CW mode, touch and hold the BKIN Soft Key to adjust the delay. Turn the Multi knob to change the setting. The default is a period of 7.5 dits at the selected KEY SPEED.

RELAY CLICKING NOISE ON CW MODE

The PTT switching on the 2m band uses a relay and the clicking noise is quite noticeable when using full break-in keying. The noise is tolerable when using semi break-in because the relay is not switching after every Morse code dit or dah symbol. If you are not a CW Whizz Kid, use semi break-in for CW on the 2 m band.

The PTT switching on the 70 cm band is silent even on full break-in because the radio uses pin diodes for the send/receive switching. On the 23 cm band, the PTT switching is reasonably quiet, because a micro-relay is used.

OVF ALERT

The OVF alert is displayed the right of the FIL (filter) indicator on each VFO. It indicates that the receiver ADC is being overloaded. Turn off the preamplifier and if necessary, turn on the attenuator using the P.AMP ATT button. Or you can turn down the RF gain by turning the RF/SQL knob anti-clockwise.

Note that the signal that is causing the overload may not be the signal that you are listening to and may not be visible on the panadapter spectrum display. Adding attenuation using the ATT Soft Key should fix the problem.

LMT ALERT

The LMT icon is beside the red TX icon. It indicates that the transceiver has overheated while transmitting and the protection system has been activated. The TX Icon will also turn to white lettering on a black background. You will not be a transmit until the radio cools down. Touch and hold the meter scale or select <MENU> <METER> to see the temperature meter.

PROBLEMS TRANSMITTING DIGITAL MODES

The audio signal is sent over the USB cable to the PC in any mode. So, you can use your digital mode PC software to receive and decode digital mode signals like RTTY, PSK, or FT8.

But you should not transmit audio digital mode signals unless you are in a DATA mode, usually USB-D. FM-D would be used for FM Packet radio. Note that the radio does not currently support 9600 baud FM Packet. The default setting for 'Data Off Mod' stops you transmitting audio digital mode signals from the PC unless you are in a DATA mode. But you can bypass that restriction if needed.

CW is digitally keyed, so an external program can key CW when the transceiver is in CW mode. AFSK RTTY is transmitted using the SSB-D mode. FSK RTTY is transmitted with the radio in RTTY mode.

VOICE KEYER, CW, OR RTTY MESSAGE KEYER NOT WORKING

Check that the message keyer matches the mode. They look similar, but the wrong one won't work. Check the text above the eight message buttons. VOICE TX for voice modes, KEYER for CW, and RTTY DECODE for RTTY.

The Soft Keys are T1-T8 for Voice, M1-M8 for CW, and RT1-RT8 for RTTY.

HELP, I SET MY RADIO TO JAPANESE TEXT MODE

It is more difficult to set the IC-9700 to Japanese text mode than it is in the IC-7300 or the IC-7610. You have to change two menu settings to achieve it. So, it is unlikely to happen by accident. However, I have heard of a user who performed a full reset on the radio and restarted in Japanese text mode.

Select <MENU> <SET> Then touch the ディスプレイ設定 'Display' icon on page 2/3. It is the row with a picture of a computer screen on it.

Then select 表示言語 'Display Language' It is the bottom menu item on page 5.

Then select 'English' 英語. It is the top menu item of the two options.

Finally, on page 6, select <System Language> <English>. This removes the Display Language option and any possibility of setting the radio back to Japanese again, other than by performing a full reset.

KEYBOARD AND COMPUTER MOUSE SUPPORT

Some forum users have been wondering how to connect an external mouse or keyboard to the IC-9700 and the sad fact is that you can't. The radio only has one USB port and it is a type B port.

A USB Type B port is used when the device, in this case, the transceiver, is a peripheral connected to a remote host, the PC. It is used on items like scanners or printers. A USB Type-A port is a host port used for connecting a USB peripheral such as a keyboard, a mouse or a USB memory stick.

Unlike the IC-7610, the IC-9700 does not have a USB Type-A port, so you cannot connect a mouse or a keyboard. This is a big problem if you want to use the built-in RTTY functions and to a lesser degree the CW message functions. The problem is that there is no way to input the callsign of the station you are working or giving a signal report to. You can't type a message either. It makes the built-in RTTY function 90% useless. I guess you could edit the RTTY message using the on-screen keyboard each time you want to send someone's callsign, but I don't think that is very practical during a QSO.

Perhaps you could edit the message and then call the station. Unfortunately, there is no easy way to fix the problem because the radio simply does not have the required USB port. Icom could easily provide an onscreen keyboard for the RTTY and CW modes. Let's hope it is included in a future firmware release.

CLOCK LOSES TIME OR RESETS

The IC-9700 has a rechargeable 3 Volt battery to keep the clock working while the radio is disconnected from the 13.8 V DC supply. If you leave the radio disconnected for an extended period, the battery will discharge, and the clock will reset. If the voltage gets too low the battery will be damaged and require replacement which is a solder job as it is not in a holder. Note that if the radio is left connected to your DC supply the clock battery will charge even if the radio is turned off. The obvious solution is to leave your DC supply running and turn the rig off with the power button. Apparently, it can take up to two days to charge from a fully discharged state.

PANADAPTER GAIN

This really bugs me. You can't change the gain of the panadapter band scope display. You can change the reference level, which shifts the whole panadapter up or down the screen, but the gain is always set to 10dB per division. This means that a lot of the time the panadapter will display tiny signal peaks just above the noise floor. A control to adjust the panadapter gain or even a Soft Key on the EXPD/SET to set a choice of 5 dB or 10 dB per division would be a big improvement.

LOW AVERAGE TRANSMIT LEVEL ON SSB

This is a complaint made by a few users of the IC-7300 and the IC-9700 transceivers. Firstly, let me point out that the LED power meter on the radio is much faster than an analog power meter so it may appear to indicate a higher average power reading than an external analog power meter.

However, the radio will output low average power output if it is not set up properly. I believe that the problem arises if you set the microphone level too high. When the user tries to fix the problem by increasing the microphone gain, they end up making the situation worse. If the microphone gain is too high, the ALC meter reading becomes compressed up into the top part of the red line meter section. The ALC circuit reduces the modulation stage gain to cope with the high level of modulating audio. It incorporates a time constant which has the effect of leaving the level low when the modulation level decreases, so while the voice peaks may make full power, the average power is reduced. The ALC is an audio stage level control, not a compressor.

If you are concerned about low average power output, try setting up the radio for SSB following the instructions on page 16. Reduce the microphone gain so that your voice just peaks the transmitter to full power and the blue ALC meter traverses the full range of the red line meter scale and is not constantly up the high end. I ended up with the microphone set at 24% but your setting may be different.

WSJT-X RIG FAILURE - PROTOCOL ERROR

You can fix a *'Rig Failure - Protocol error while getting current VFO frequency'* error message from WSJT-X by setting <MENU> <SET> <Connectors> <CI-V> <CI-V USB Echo Back> to <ON>.

This seems to have been fixed in the latest release of WSJT-X. So if you encounter this error you should update to the latest version of WSJT-X.

Technical Notes

OSCILLATOR STABILITY

There has been a great deal of Internet forum traffic and a couple of videos discussing the frequency stability of the IC-9700. Much of the consternation and hysteria stems from the inclusion on the back panel of a 10 MHz reference signal input. Initially, this input could only be used to calibrate the internal oscillator, but that was changed with firmware release 1.10 in June 2019. Now the 10 MHz reference signal input can be used to govern the internal master clock in the radio.

The radio's main clock oscillator has the same frequency stability as other recent Icom models including the IC-9100, IC-7100, ID-5100, IC-R8600, and IC-7610. The IC-9700 is no better and no worse than any other model. The amount of frequency drift, although quite rapid after a transmit period, is well within the published ±0.5 ppm specification for the transceiver. For basic DV, FM, and SSB operation the stability of the internal clock is totally acceptable on any of the bands. However, if you are using the radio for EME work, WSPR, FT8, or other narrowband digital modes on the 70 cm or 23 cm band, the new firmware revision is very welcome because the ability to frequency lock the radio is essential.

The radio is very stable while receiving. After the receiver had been running for 30 minutes, I measured a frequency error of 13 Hz at 1296.2 MHz which is outstanding. The problem arises when the transceiver is cooling down following a transmit period. The amount of radio frequency drift is very dependent on the temperature inside the radio. After transmitting the fan acts to cool down the radio and this accelerates the frequency drift during the WSPR or JT65 receive period. The amount of drift is significant on the 70 cm and 23 cm bands and it will definitely be a problem when using very narrow bandwidth digital modes such as WSPR and JT65 on those bands. After a one minute transmit period, I measured a drift of around 30 Hz on the 70 cm band over the one-minute receiving period. This amount of drift is not a big problem for SSB or FM operation, but it can cause poor decodes on JT65 mode. Using a GPS reference clock, the drift after a transmit period is within 5 Hz. If clock stability is still a problem for you, there is a hardware modification available.

The 10 MHz reference input is used to apply a very stable reference clock, usually a GPS referenced clock or a Rubidium frequency standard, that will lock, or 'govern' the radio's main oscillator, vastly improving the frequency stability and reducing the amount of frequency drift to almost zero.

This feature is also included on several other Icom models including the IC-7610, the IC-R6800 receiver, and the top end Icom transceivers such as the IC-7851.

Some people use the term GPSDO (GPS disciplined oscillator) when referring to GPS referenced clocks. I prefer to use the term 'clock' because a clock has a square wave (or pulse train) output. A GPSDO could have a sine wave or a square wave output. You can apply either a sine wave or a square wave to the reference input connector.

I carried out a conversation on SSB at 1296.2 MHz and you could certainly tell that the frequency was drifting. I had to compensate by using the RIT control, but that was no problem. The table below shows the acceptable frequency offset for a radio with a ±0.5ppm TXCO. I am not aware of any measurement done on the radio that indicates a level of frequency drift that even comes close to exceeding this standard. I am also not aware of any unmodified commercial ham band VHF/UHF transceiver that exceeds this level of stability without an external reference.

Expected maximum frequency drift for a ±0.5 ppm TXCO		
Band	VFO frequency	Max frequency error
2 m	146 MHz	VFO ± 73 Hz
70 cm	430 MHz	VFO ± 215 Hz
23 cm	1296 MHz	VFO ± 648 Hz

If you do apply a GPS referenced or an atomic clock, the frequency stability while receiving should be at least as good as the figures in the following table and may be ten to one hundred times better than the numbers stated. You will still see a small (5 to 15 Hz) amount of frequency instability in the 60 seconds following a transmit period as the software control loop acts to control the oscillator frequency.

Expected maximum frequency drift for a GPS reference frequency locked radio, (±1 x 10^{-9} stability). (Long term receiving)		
Band	VFO frequency	Max frequency error
2 m	146 MHz	VFO ± 0.14 Hz
70 cm	430 MHz	VFO ± 0.43 Hz
23 cm	1296 MHz	VFO ± 1.30 Hz

Oscillator stability is measured in ppm (parts per million). If the main oscillator or clock is off frequency, all of the higher frequencies that are generated from that master clock or oscillator will be off by the same ratio. The quoted stability for the TXCO (temperature-controlled oscillator) in the IC-9700 is plus or minus 0.5 ppm. That means at 430 MHz the *frequency could be plus or minus 215 Hz. Therefore, the maximum frequency error is 430 Hz. A GPS referenced clock uses the very accurate time codes from GPS satellites to control the drift of the TXCO. A GPS referenced clock can achieve frequency stability better than 1 ppb ($1x10^{-9}$) which is about 500 times better than the TXCO on its own. It could hold the transceiver to less than ±1 Hz of error even on the 23 cm band. Many GPS referenced clocks can approach a stability of 1ppt ($1x10^{-12}$) which is about 500,000 times better than the TXCO.*

APPLYING A GPS REFERENCE CLOCK TO THE 10 MHZ REFERENCE

If you want to calibrate the radios internal TXCO or govern the TXCO to improve the radio's frequency stability, you can apply a 10 MHz GPS reference clock or a Rubidium clock or oscillator to the 10 MHz reference input. The previous section explains about oscillator stability and the reasons why you may want to connect a super stable oscillator or clock to the reference input.

For frequency locking, your radio must be running Firmware version 1.10 or newer. For frequency calibration only, your radio must be running Firmware version 1.06 or newer.

➢ Requirements

The nominal input level for the 10 MHz reference input is -10 dBm. The level is not supercritical. There is a diode limiter in the circuit. You can safely use levels ranging from -5 dBm to -25 dBm. Use an inline 50 Ohm coaxial attenuator if you need to reduce the output level of your reference signal.

The reference input signal needs to be very stable. There is no point in connecting a 10 MHz source that is less stable than the radios ±0.5 ppm TXCO.

The reference clock or oscillator also needs to have very low phase noise and jitter. Receiver and transmitter performance may be affected if you connect a poor-quality signal source.

The input is designed for a clock, (square wave signal), but you will still get a lock using a 10 MHz sinewave GPSDO (GPS disciplined oscillator). The signal will be squared and limited within the radio.

➢ Leo Bodnar GPS Reference Clock

I carried out some experiments locking the IC-9700 to a Leo Bodnar GPS Reference Clock. These devices are very stable and reasonably priced. The clock can be set to exactly 10.000000 MHz and its stability is better than 1×1^{-10}. The long-term stability is better than 1×10^{-11}.

Once the frequency has been set using the free software, you can disconnect the unit from the computer and run it from a 5 – 14 V DC supply or a USB power supply. It will remember the frequency setting even if the power is removed. If you are intending to use the radio for narrowband digital modes such as JT65 on the 70 cm or 23 cm bands, a GPS reference clock is a good idea. The same is true if you are using the radio with microwave band transverters or converters. Depending on the frequency required, the two-port model could be used to frequency lock both the radio and the LNB. However, the choice of second channel frequencies is quite limited.

The Leo Bodnar GPS Reference Clock outputs about +7.4 dBm when set to the 8mA output and about +11.5 dBm when set to the 16 mA output. This is a bit too high for the -10 dBm reference input on the IC-9700. The radio will sync to a -22 dBm input signal with no problem, so I use the 8mA output on the GPS clock with a 20 dB 50 Ohm coaxial attenuator between the GPS clock and the radio and it works fine.

➢ Settings

Connect the 10 MHz clock to the 10 MHz reference input. It is an SMA connector. I use a SMA to BC adapter because my attenuator and GPS reference clock are fitted with BNC connectors.

To start the calibration or establish frequency locking, use <MENU> <SET> <Function> <REF Adjust> <Sync to REF IN> <START>. After a message, the radio will synchronize the master oscillator to the incoming reference signal.

You may see some changes in the coarse and fine frequency numbers. That is the calibration working. Don't change these settings back to what they were after you remove the reference signal. If you press <Cancel Sync> or remove the reference input the radio will now be calibrated.

While the reference signal is connected it will govern the master TXCO improving the overall stability dramatically. The GPS reference clock should be able to hold the radio to better than ±1.5 Hz on the 23 cm band and better than ±1 Hz on 70 cm and 2m. The improved stability should stop any frequency drift while transmitting and it will hold the receiver frequency to a reasonable (5 -15 Hz) amount of drift while the radio is cooling down after transmitting. After about a minute, the receiver frequency will have returned to a high stability state.

If you press <Cancel Sync> or remove the reference input, the radio will remain calibrated but the oscillator stability will return to ±0.5 ppm which is fine for general operation.

THE ADC

The IC-9700 uses a 14-bit LTC2156-14 dual ADC (analog to digital converter). One ADC is used to receive either the 2 m band or the 23 cm band. The other is used to receive either the 70 cm band or the 23 cm band. In that way, the radio can receive on any two of the three bands simultaneously.

The LTC2156-14 has a maximum sample rate of 210 Msps and a sampling bandwidth up to 900 MHz. The ADC sample rate is 196.608 Msps which is four times the 49.152 MHz Main Oscillator frequency. This sampling frequency allows the three bands to fall neatly into individual Nyquist zones.

The 144-148 MHz (2 m band) falls into the 2nd Nyquist zone (98.304-196.608 MHz)

The 311-371 MHz (23 cm I.F.) falls into the 4th Nyquist zone (294.912-393.216 MHz)

The 430-440 MHz (70 cm band) falls into the 5th Nyquist zone (393.216-491.52 MHz)

The ADC uses internal pipelining to average out quantization errors. This helps to achieve a high spurious free dynamic range (SFDR). It also includes an optional 'randomization' function to improve intermodulation performance.

This function is turned on with the IP+ function in the radio. The randomization function works in the usual way by applying an 'Exclusive Or' logic function to bits 1 to 13 based on the value of bit 0. There is also an 'alternate bit polarity reversal' function that reverses the polarity of all the odd bits, (1, 3, 5, 7, 9, 11, and 13) so that the overall bitstream is more random. Apparently, this reduces digital noise getting back into the input signal especially when the input signal level is very low. The reversal would have to be changed back at some stage, but that could be done very easily inside the FPGA software. I don't know if Icom is using this option. The ADC does not include a 'dither' option.

Rather unusually there are only seven output pins for each ADC. When the clock goes low, the device outputs the value of bits 0, 2, 4, 6, 8, 10, and 12, then when the clock signal goes high, it outputs bits 1, 3, 5, 7, 9, 11, and 13. This is a rather neat way of outputting 14 bits of data per clock cycle while minimizing the number of output pins required.

ADC SIGNAL TO NOISE RATIO

The ADC SNR (Signal to noise ratio) is included in the LTC2156 datasheet.

SNR @ 150 MHz = 68 dB (121 dB including process gain to a 500 Hz bandwidth). SNR @ 450 MHz = 66 dB (119 dB including process gain to a 500 Hz bandwidth).

The actual MDS (minimum discernible signal) in a 500 Hz or 2400 Hz bandwidth will be affected by any preamplifier gain and/or attenuation before the ADC.

IC-9700 SENSITIVITY (MDS)

The sensitivity numbers listed below are from the Icom Basic manual. Independent lab reports indicate that the radio easily exceeds the quoted sensitivity figures. Note that the data provided by Icom is for a test done with the preamplifier on and also with IP+ on (independent lab tests usually use a 500 Hz bandwidth and have the preamplifier turned off). The band that was tested is not stated. Assuming that the supplied data is for a 500 Hz bandwidth, the calculated MDS figure is -136.2 dBm.

Sensitivity	Input level	Preamp on and IP+ on	Input dBm	MDS dBm
SSB	0.11 uV pd	for 10dB SNR	-126.2	-136.2
FM	0.18 uV pd	for 12 dB SINAD	-121.9	
AM	1 uV pd	for 10dB SNR	-107.0	
DV	0.35 uV pd	For 1% BER	-116.1	
DD (23 cm)	1.59 uV pd	For 1% BER	-103.0	

23 CM CONVERTER

The 23 cm down converter uses a local oscillator on 929 MHz to convert the 23 cm band of 1240 – 1300 MHz down to 311 – 371 MHz which is in a range that the ADC can sample.

The 23 cm band transmit signal comes out of the DAC on a frequency between 305.632 MHz and 365.632 MHz. Prior to the driver stage, it is upconverted to the wanted frequency on the 23 cm band by using a mixer and fixed oscillator on 1605.632 MHz, before being applied to the input to the 10-Watt power amplifier.

Modifications

Please note that I am not endorsing any modifications to the radio and that carrying out any hardware modification may void the Icom warranty.

"This warranty shall not apply: To an Icom product which has failed to function as required due to improper installation, misuse, accident, alteration or unauthorized repair or modification." [RWB Communications website – NZ Icom agent].

TXCO STABILITY UPGRADES

➢ Mini-Kits GPS-9700

Mini-Kits https://www.minikits.com.au/ offers a GPS locked 49.152 MHz signal source that governs the internal TXCO. The device is offered as a kit, a built-up board, or a fully assembled unit in a small Hammond case. The design incorporates a 0.28ppm low phase noise 49.152MHz Voltage Controlled Temperature Controlled Crystal Oscillator (VCTCXO) reference. The frequency is amplified and applied to the IC-9700's internal 49.152MHz frequency reference using a small coupler PC board to mode lock the frequency. The GPS-9700 module can be used with or without GPS locking and is much more stable on its own compared to the IC- 9700 due to it being separated from temperature changes within the Transceiver. For high-frequency stability narrow-band weak signal modes, a high-quality low phase noise external 10MHz GPS source is required. There are no modifications required to the Icom IC-9700, but a small coupler PC board with a single solder connection and replacement cable is required to be fitted internally inside the Transceiver.

This is a relatively low impact modification as the only soldering required is to mount a small board inside the transceiver. The circuit works by passive coupling into the radio's oscillator circuit rather than a direct connection.

Using the Mini-Kits GPS-9700 without the connection of an external 10 MHz source, the stability of the radio is improved from 0.5 ppm to 0.28 ppm. While this seems insignificant, the real bonus is that the oscillator resides outside of the radio and is isolated from the temperature changes caused by transmitting which are the chief source of the problem. This may be all the improvement that you need.

With the addition of a 10 MHz frequency standard, the radio stability will be similar to connecting a 10 MHz frequency standard directly to the Reference Input connector, except again, the stability will be less affected by temperature changes in the radio.

My advice would be to try a 10 MHz GPS referenced clock connected to the Reference Input connector first. In the event that you are still experiencing problems with JT65 or your microwave transverter, or you want a hardware solution, then consider adding the Mini-Kits GPS-9700 device as well. It is a shame that this unit does not include a GPS receiver to supply the 10 MHz reference signal so that you wouldn't have to buy both devices.

➢   VK1XX oscillator upgrade

Glen English VK1XX has also developed a hardware GPS locking solution. Fitting the board requires soldering the board to one of the metal oscillator covers and soldering two short wires to the radio PCB. The board will sense when it has a 10 MHz signal from the 10 MHz GPS referenced clock input and will return control to the Icom TXCO if the GPS signal if it is not connected. You can turn the 'Sync' on and off in the same way as just using the GPS referenced clock without the add-on board. As at August 2019, these boards are not yet commercially available. But a test batch of 80 boards has been commissioned.

https://www.youtube.com/watch?v=dp89jN2Fo28

http://www.cortexrf.com.au/IC9700lock.html

FAN MODIFICATION

When users were first investigating the oscillator stability issue, it was found that having the fan running at a moderate speed continuously was better than the normal Icom fan switching. The modification consists of adding a 6.8V 5W Zener diode across the transistor that switches the fan on and off. This modification is no longer required because you will get a much better improvement by adding a 10 MHz GPS referenced clock and turning on the Icom Synchronization.

FREQUENCY MODIFICATION

This modification makes the frequency range of a European version of the radio the same as an American or International version.

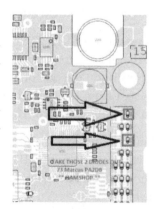

Removal of D332 and D326 expands the 2m band from 144-146 MHz to 144-148 MHz and extends the 70cm band from 430-440 MHz out to 430-450 MHz.

My radio already has those settings.

Glossary

Term	Description
2 m, 70 cm 23 cm	144-148 MHz, 430 - 450 MHz, and 1240 - 1300 MHz amateur radio bands
59	Standard (default) signal report for amateur radio voice conversations. A report of '59' means perfect readability and strength.
599, 5NN	Standard (default) signal report for amateur radio CW conversations. A 599 report means perfect readability, strength, and tone. The 599 signal report is often used for digital modes as well. The 5NN version is faster to send using CW. It is often used as a signal report exchange when working contest stations.
73	Morse code abbreviation 'best wishes, see you later.' It is used when you have finished transmitting at the end of the conversation.
.dll	Dynamic Link Library. A reusable software block which can be called from other programs.
A/D	Analog to digital
ACC	Accessory jack
ADC	Analog to digital converter or analog to digital conversion
AF	Audio frequency - nominally 20 to 20,000 Hz.
AFSK	Audio Frequency Shift Keying. A digital mode that uses tones rather than a digital signal to drive the SSB transmitter.
AGC	Automatic Gain Control. In an SDR like the IC-9700, it limits the receiver audio output in the presence of large signals.
ALC	Automatic Level Control. There are two kinds used by the radio. One is the ALC used to ensure that the radio is not overmodulated. It is metered by the ALC meter. The other is the ALC control from a linear amplifier which ensures that it is never overdriven by the transceiver.
Algorithm	A process, or set of rules, to be followed in calculations or other problem-solving operations, especially by a computer. In DSP it is a mathematical formula, code block, or process that acts on the data signal stream to perform a particular function, for example, a noise filter.
AM	Amplitude modulation, (double sideband with carrier)

ANF	The Automatic Notch Filter eliminates the effect of long-term interference signals such as carrier signals that are close to the wanted receiving frequency. Not effective against impulse noise.
ANT	Abbreviation for antenna
ATT	Abbreviation for attenuator
Auto Tune	This is a button which pulls the radio VFO on to the frequency that matches the CW Pitch that you have set. It 'Nets' the CW receive frequency and transmit frequency so that you will transmit CW on the same frequency as you are receiving. It only works in the CW mode.
Band scope	A band scope is a spectrum display of the frequencies above and below the frequency that the radio is tuned to. The center of the display is generally the frequency that you are listening to. This is different from a panadapter where you can listen to any frequency across the display.
Bit	Binary value 0 or 1.
BKIN or F-BKIN	CW 'Break-in' the practice of receiving Morse code between the Morse characters or words that you are sending. The radio will not automatically transmit in the CW mode unless BKIN is set to either semi break-in BKIN or full break-in F-BKIN.
BPSK	Binary phase-shift keying. Digital transmission mode using a 180-degree phase change to indicate the transition from a binary one to a binary zero. The number is the baud rate. BPSK31 is slower but uses less bandwidth and is easier to decode than BPSK63.
Carrier	Usually refers to the transmission of an unmodulated RF signal. It is called a carrier because the modulation process modifies the un-modulated RF signal to carry the modulation information. A carrier signal can be amplitude, frequency, and/or phase-modulated. Then it is referred to as a 'modulated carrier.' An oscillator signal is not a carrier unless it is transmitted.
CAT	Computer-aided transceiver. Text strings used to control a ham radio transceiver from a computer program. The Icom standard for CAT control is called CI-V
CENT	Center mode. Sets the panadapter display so that the main VFO frequency is in the center of the display with a SPAN of frequencies either side.
CI-V	Icom standard for CAT control. Text strings used to control a ham radio transceiver from a computer program
CODEC	Coder/decoder - a device or software used for encoding or decoding a digital data stream.

COM	Serial Communications port. In the IC-9700 each Com Port is a 'virtual' serial port carried over the USB cable. The dedicated CI-V 'REMOTE' connector is also a COM port but a level converter is required.
COMP	Compressor. Increases the average power of your transmission by decreasing the dynamic range of the audio signal. i.e. makes the quiet parts louder.
CPU	Central processing unit - usually a microprocessor. Can be implemented within an FPGA.
CQ	"Seek You" an abbreviation used by amateur radio operators when making a general call which anyone can answer.
CTCSS	Continuous tone coded squelch system. A sub-audible tone that is sent with your transmission to open the receiver squelch at a repeater. May also be transmitted by the repeater to perform the same function in your receiver.
CW	Continuous Wave. The mode used to send Morse Code.
D/A	Digital to analog.
DAC	Digital to analog converter or digital to analog conversion
DATA	One of the data modes (AM-D, FM-D or SSB-D) used to interface the radio with a PC digital mode program. You must be in a DATA mode to transmit from a PC digital mode program.
dB, dBm, dBc, dBV	The Decibel is a way of representing numbers using a logarithmic scale. dB is used to describe a ratio, i.e. the difference between two levels or numbers. Decibels are often referenced to a fixed value such as a Volt (dBV), a milliwatt (dBm), or the carrier level (dBc). Decibels are also used to represent logarithmic units of gain or loss. An amplifier might have 3 dB gain. An attenuator might have 10 dB loss.
DC	Direct Current. You need an 18 Amp regulated 13.8V DC supply to power the radio.
DD	Digital data. On the 23 cm band, the IC-9700 can transmit data from an external source such as the Internet, a PC, or a GPS receiver, at a rate of up to 128 kbits per second. It can also receive data and send it back to the PC or Internet connection. For example, using the radio connected to a PC, you could browse the internet through a repeater gateway connection.
Digital modes	Amateur radio transmission of digital information rather than voice. It can be text, or data such as video, still pictures, or computer files, (PSK, RTTY, FT8, Olivia, SSTV, etc.)

Digital voice	Amateur radio mode where speech is coded into a digital format and sent as tone sequences or phase shift keying. In the IC-9700 it means D-Star.
DR3	3rd order dynamic range receiver test
DSP	Digital signal processing – a dedicated integrated circuit chip usually running internal firmware code, or a software program running on a computer. DSP uses mathematical algorithms in computer software to manipulate digital signals in ways that are equivalent to functions performed on analog signals by hardware mixers, oscillators, filters, amplifiers, attenuators, modulators, or demodulators.
DTR	'Device terminal Ready' a com port control line often used for sending CW or FSK RTTY data over the CI-V interface between the radio and a PC.
DV	Digital voice mode. In this case, it refers to D-Star.
DX	Long-distance, or rare, or wanted by you, amateur radio station
DXCC	The DX Century Club. An awards program based around confirming contacts with 100 DXCC 'countries' or 'entities' on various modes and bands. The DXCC list of 304 currently acceptable DXCC entities is used as the worldwide standard for what is a separate country or recognizably separate island or geographic region.
DXpedition	A DXpedition is a single, or group of, amateur radio operators who travel to a rare or difficult to contact location, for the purpose of making contacts with as many amateur radio operators as possible worldwide. They often activate rare DXCC entities or islands and they may operate stations on a number of bands and modes simultaneously.
EXPD	Abbreviation for Expanded
EXT	Abbreviation for External
FFT	Fast Fourier Transformation – conversion of signals from the time domain to the frequency domain (and back using IFFT).
FIX	Fixed mode. Sets the panadapter display so that it displays the range frequencies between two pre-set frequencies determined by Band Edges. The VFO frequency is indicated by the green marker.
FM	Frequency Modulation.
FPGA	Field Programmable Gate Array – a chip that can be programmed to act like logic circuits, memory, or a CPU.
FSK	Frequency Shift Keying.

FSK RTTY	FSK RTTY is keyed using a digital signal to offset the transmit frequency rather than AFSK which generates audio tones at the Mark and Space offsets from the VFO frequency.
GND	Ground. The earthing terminal for the radio. This should be connected to a 'telecommunications' ground spike, not the mains earth.
GPS	Global Positioning System. A network of satellites used for navigation, location, and very accurate time signals.
GPS referenced clock	GPS disciplined oscillator. An oscillator locked to time signals received from GPS satellites
Hex	Hexadecimal is a base 16 number system used as a convenient way to represent binary numbers. The default Icom address for the IC-9700 is A2h (162 decimal) (1010 0010 binary)
HF	High Frequency (3 MHz -30 MHz)
Hz	Hertz is a unit of frequency. 1 Hz = 1 cycle per second.
IF or I.F.	Intermediate frequency = the Signal – LO (or Signal + LO) output of a mixer
IMD	Intermodulation distortion. Interference or distortion caused by non-linear devices like mixers. There are IMD tests for receivers and transmitters. IMD performance of linear amplifiers can also be tested.
IP+	The IP+ control enables ADC randomization in order to optimize the sampling system for best receive IMD (intermodulation distortion) performance. It can improve receiver performance in the presence of very large interfering signals. But there is a small loss of sensitivity. While it improves laboratory 'two tone IMD test' results, it is usually unnecessary in 'on-air' situations.
IQ	Refers to the I and Q data streams treated as a pair of signals. For example, a digital signal carrying both the I (incident) and Q (quadrature) data.
Key	A straight key, paddle, or bug, used to send Morse Code
kHz	A kilohertz is a unit of frequency. 1 kHz = 1 thousand cycles per second.
LAN	Local Area Network. The Ethernet and WIFI connected devices connected to an ADSL or fiber router at your house is a LAN.
LED	Light Emitting Diode
LSB	Lower sideband SSB transmission

m, 2 m, 6m	Meter (US) or Meter. Often used to denote an amateur radio or shortwave band; e.g. 2 m, 6 m, 30 m, 10 m, where it denotes the approximate free-space wavelength of the radio frequency. Wavelength = 300 / frequency in MHz. A frequency range of 3 to 30 MHz has a corresponding wavelength of 100m to 10m.
Marker	The R and T markers indicate the frequency of the receiver (in FIX mode) and the transmitter frequency (press XFC).
MDS	Minimum discernible signal. A measurement of receiver sensitivity
Menu	The MENU button displays a wide range of settings that are not used enough to warrant a dedicated front panel button.
MHz	Megahertz – unit of frequency = 1 million cycles per second.
MIC	Microphone
MPAD	Memory Pad. A short-term memory function, storing either five or ten frequency and mode settings
MW	Memory Write Soft Key
NB	Noise Blanker. A filter used to eliminate impulse noise
Net	An on-air meeting of a group of amateur operators. Or to 'Net' the CW receive frequency and transmit frequency so that you will transmit CW on the same frequency as you are receiving
NR	Noise Reduction. A filter used to eliminate continuous background noise
Onboard	A feature performed within the radio. Especially one that usually requires external software. For example, the radio has an 'onboard' RTTY decoder.
OVF	Overflow. The overflow warning indicates that the receiver ADC is being overloaded with very big received signals. This will cause problems on the panadapter display and possibly create noise due to severe intermodulation distortion. Turn on the attenuation with the ATT button or reduce the RF Gain (RF/SQL control).
P.AMP	An abbreviation for 'preamplifier'
Panadapter	Panadapter is short for Panoramic Adapter. It allows us to see a panoramic display of the band. The IC-9700 panadapter can display a spectrum display and optionally a waterfall picture. You may be listening to one or more signals anywhere within the displayed spectrum of frequencies. This is different from a band scope.
PC	Personal Computer. For the examples throughout this book, it means a computer running Windows 10.

Pileup	A pileup is a situation when a large number of stations are trying to work a single station, for example, a DXpedition or a rare DXCC entity. Split operation is often employed to spread the pileup of calling stations over a range of frequencies.
Po	RF power output (meter)
PROG	PROG (Programmed) is a Soft Key used to scan between the pre-set P1 and P2 frequencies.
PSK	Phase shift keying. Digital transmission mode using phase change to indicate the transition from a binary one to a binary zero.
PTT	Press to talk - the transmit button on a microphone – The PTT signal sets the radio and software to transmit mode. Icom calls it 'SEND.'
QPSK	Quadrature phase-shift keying. Digital transmission mode using 90-degree phase changes to indicate four two-bit binary states 00,01,10,11.
QRP	Q code - low power operation (usually less than 10 Watts).
QSO	Q code – an amateur radio conversation or "contact."
QSK	Q code – fast transmit to receive switching which allows Morse code to be received in the gaps between the CW characters that you are sending.
QSY	Q code – a request or decision to change to another frequency.
REC	Record. Used to record off-air received audio. Also, a Soft Key used to record microphone audio for the Voice Message keyer.
REF	Reference
RF	Radio Frequency
RIT	Receive Incremental Tuning. A way to fine-tune the signal that you are receiving without changing the main VFO and hence your transmitted frequency. Used when the other station is a little off frequency.
RS232	A computer interface used for serial data communications.
RTTY	Radio Teletype. RTTY is a frequency shift, digital mode. Characters are sent using sequences of Mark and Space tones.
RTS	'Ready to Send' a com port control line often used for sending the PTT (SEND) command over the CI-V interface between the radio and a PC.
RX	Abbreviation for receiver
SDR	Software Defined Radio. Actually, the IC-9700 is more correctly a 'direct sampling' radio.

Sked	A pre-organized or scheduled appointment to communicate with another amateur radio operator
SNR	Signal-to-Noise Ratio in dB (decibels).
Soft Key	A button or selectable icon displayed on the touch screen
Split	The practice of transmitting on a different frequency to the one that you are receiving on. Split operation is commonly used by DXpeditions and anyone who generates a large pileup of callers. SSB split is commonly 5-10 kHz. CW split is commonly 1-2 kHz.
Squelch	Squelch mutes the audio to the speakers when you are not receiving a wanted signal. When the received signal level increases the squelch opens and you can hear the station. Squelch does not affect the audio output over the USB cable or the audio meter display, (menu setting).
SSB	Single sideband transmission mode.
SWR	Standing Wave Ratio. The RF power reflected back from a mismatched antenna or connection.
TU	Morse code abbreviation meaning 'to you.' It is used when you have finished transmitting and wish the other station to respond.
TX	Abbreviation for Transmit or Transmitter
UHF	Ultra High Frequency (300 MHz - 3000 MHz).
USB	Universal serial bus – serial data communications between a computer and other devices. USB 2.0 is fast. USB 3.0 is very fast.
USB	Upper sideband SSB transmission.
UTC	Universal coordinated time. UTC is the standard time used by amateur radio operators. Everyone logging contacts using the same UTC time rather than local time makes comparing logs and confirming contacts much easier.
VBW	Video Bandwidth is the ability of the spectrum scope (panadapter) to distinguish weak signals from noise. A narrow VBW can filter noise but requires more processing power.
VFO	Variable Frequency Oscillator. The IC-9700 has two VFOs called 'A' and 'B.' The main tuning knob controls the active VFO.
VHF	Very High Frequency (30 MHz -300 MHz)
VOX	Voice Operated Switch. Voice-activated receive to transmit switching
W	Watts – unit of power (electrical or RF).

Index

References and Links

1. Icom IC-9700 Basic Manual (copyright Icom) and Icom IC-9700 Full Manual (copyright Icom) http://www.icom.co.jp/world/support/download/manual/disp_products.php?SWT=&INQWORDSWT=Yes

2. Icom Firmware http://www.icom.co.jp/world/support/download/firm/index.html.

3. Icom repeater list http://www.icom.co.jp/world/support/download/firm/IC-9700/repeater_list/.

4. Icom CS-9700 programming software https://www.icom.co.jp/world/support/download/firm/IC-9700/CS-9700_1_01/

5. There is an excellent video on setting up the Icom RS-BA1 remote control software at https://www.youtube.com/watch?v=vhhNo2AoO0Y

6. Radio Society of Great Britain www.rsgb.org

7. N1MM logger using MMTTY, refer to the notes by K0PIR at http://www.k0pir.us/icom-7300-rtty-fsk-mmtty/ or a video on the topic at https://www.youtube.com/watch?v=NmNHVjjAdiY

8. Leo Bodnar GPS referenced clock http://www.sdr-kits.net/ or http://www.leobodnar.com/shop/index.php?main_page=index&cPath=107&zenid=7a6ab92cd46db8b3e49cd2019b578403

9. VK1XX oscillator modification https://www.youtube.com/watch?v=dp89jN2Fo28 http://www.cortexrf.com.au/IC9700lock.html

10. Mini-Kits GPS-9700 https://www.minikits.com.au/

11. SatPC32 is available as a download from http://www.dk1tb.de/indexeng.htm and also from the AMSAT-NA, AMSAT-UK, and AMSAT-DL websites.

12. D-Star reflectors http://www.dstarinfo.com/Reflectors.aspx.

The Author

Well, if you have managed to get this far you deserve a cup of tea and a chocolate biscuit. It is not easy digesting large chunks of technical information. It is probably better to use the book as a technical reference. Anyway, I hope you enjoyed the book and that it has made life with the IC-9700 a little easier.

I live in Christchurch, New Zealand. I am married to Carol who is very understanding and tolerant of my obsession with amateur radio. She describes my efforts as "Andrew playing around with radios." We have two children and some tropical fish. Sadly, the cat has died of old age. One son has recently graduated from Canterbury University with a degree in Commerce and the other has just graduated as a doctor after studying Medicine at Otago University.

I am a keen amateur radio operator who enjoys radio contesting, chasing DX, digital modes, and satellite operating. But I am rubbish at sending and receiving Morse code. I write extensively about many aspects of the amateur radio hobby, writing occasional columns for several magazines and other publications. This is my sixth amateur radio book.

Thanks for reading my book!

73 de Andrew ZL3DW.

THE END

73 and GD DX

Quick Reference Guide

Audio input level from PC (USB cable)	Select \<MENU\> \<SET\> \<Connectors\> \<MOD Input\> \< USB MOD Level\>.
Audio output level to PC (USB cable)	\<MENU\> \<SET\> \<Connectors\> \<USB AF/IF Output\> \<AF Output Level\>
Audio or 12 kHz IF output to USB cable	\<MENU\> \<SET\> \<Connectors\> \<USB AF/IF Output\> \<Output Select\> \<AF or IF\>
Audio input level from PC (ACC jack)	Select \<MENU\> \<SET\> \<Connectors\> \<MOD Input\> \< ACC MOD Level\>.
Audio output level to PC (ACC Jack)	\<MENU\> \<SET\> \<Connectors\> \<ACC AF/IF Output\> \<AF Output Level\>
Audio mix (Main and Sub receivers) to phones and external speakers	\<MENU\> \<SET\> \<Connectors\> \<Phones\> \<MENU\> \<SET\> \<Connectors\> \<External Speaker Separate\>
Audio scope settings	\<MENU\> \<AUDIO\> \<hold EXPD/SET\>
Band Edge (FIX panadapter)	Hold \<M.SCOPE\> Hold \<EXPD/SET\> \<Fixed Edges\> \<MENU\> \<SET\> \<Function\> \<Band Edge Beep\> \<ON User) & TX Limit\>
Band Edge (radio) setting	\<MENU\> \<SET\> \<Function\> \<User Band Edge\>
Beep settings	\<MENU\> \<SET\> \<Function\> \<Band Edge Beep\>
Call sign for DV mode	\<MENU\> \<SET\> \<My Station\> \<My Call Sign DV\>
Callsign display during power on splash screen	\<MENU\> \<SET\> \<My Station\> \<My Call Sign DV\> \<MENU\> \<Display\> \<Opening Message\> \<ON\>
CI-V settings	\<MENU\> \<SET\> \<Connectors\> \<CI-V\>
Clock – set date and time	\<MENU\> \<SET\> \<Time Set\> \<Date/Time\>
Compressor (speech)	\<FUNCTION\> \<COMP\>
CW Key / paddle type	CW mode \<MENU\> \<KEYER\> \<EDIT/SET\> \<CW-KEY SET\> \<Key Type\>
CW messages (CW mode)	CW mode \<MENU\> \<KEYER\> \<EDIT/SET\> \<EDIT\>
CW messages via keypad F1 – F4	\<MENU\> \<SET\> \<Connectors\> \<External Keypad\> \<KEYER\> \<ON\>
CW settings	CW mode \<MENU\> \<KEYER\> \<EDIT/SET\>
CW sideband LSB (default)	\<MENU\> \<SET\> \<Function\> \<CW Normal Side\>
CW sidetone (CW mode)	\<MENU\> \<KEYER\> \<EDIT/SET\> \<Sidetone Level\>
Display settings	\<MENU\> \<SET\> \<Display\>
DV/DD settings	\<MENU\> \<SET\> \<DV/DD Set\>
Echo on (for WSJT-X FT8)	\<MENU\> \<SET\> \<Connectors\> \<CI-V\> \<CI-V USB Echo Back\> \<ON\>
Firmware information	\<MENU\> \<SET\> \<Others\> \<Information\> \<Version\>

Firmware update (SD card)	\<MENU\> \<SET\> \<SD Card\> \<Firmware Update\>
	\<MENU\> \<SET\> \<Tone Control/TBW\> \<RX\> *\<mode\>*
High and Low pass filters (RX)	\<RX HPF/LPF\>
Lock (dial or panel lock)	\<MENU\> \<SET\> \<Function\> \<Lock Function\>
Log functions	\<MENU\> \<SET\> \<QSO/RX Log\>
Memory	\<MENU\> \<MEMORY\>
Memory Pad	\<MENU\> \<MPAD\>
Memory Pad size (5 or 10)	\<MENU\> \<SET\> \<Function\> \<Memo Pad Quantity\>
Message for DV mode TX	\<MENU\> \<SET\> \<My Station\> \<TX Message DV\>
Mic Gain (SSB, AM, FM)	\<MULTI\> \<Mic Gain\>
Monitor (transmit)	\<MULTI\> \<MONITOR\>
Multi-function meter	\<MENU\> \<METER\>
Mute Sub Band on TX	\<MENU\> \<Set\> \<FUNCTION\> \<Sub Band Mute (TX)\>
Network settings	\<MENU\> \<SET\> \<Network\>
Notch (auto or manual)	\<NOTCH\> or \<FUNCTION\> \<NOTCH\>
Notch width (manual)	\<hold NOTCH\> \<MULTI\> \<NOTCH WIDTH\>
	Hold \<M.SCOPE\> Hold \<EXPD/SET\> \<Fixed Edges\>
Panadapter Fixed Edges	(three per band)
Playback recorded audio file	\<MENU\> \<RECORD\> \<Play Files\>
Power on splash screen	\<MENU\> \<SET\> \<Display\> \<Opening Message\>
settings	\<MENU\> \<SET\> \<Display\> \<Power ON Check\>
	\<MENU\> \<SET\> \<Connectors\> \<External P.AMP\>
Preamplifier (external power)	*\<band\>*
	\<QUICK\> \<\<REC Start\>\>
Record an audio file	\<QUICK\> \<\<REC Stop\>\>
Reference synch to 10 MHz	\<MENU\> \<SET\> \<Function\> \<REF Adjust\>
RF gain / Squelch control	\<MENU\> \<SET\> \<Function\> \<RF/SQL Control\>
Reset (partial or full)	\<MENU\> \<SET\> \<Others\> \<Reset\>
RF Power adjust	\<MULTI\> \<RF Power\>
	\<RIT\> turn \<MULIT\>
RIT	Clear RIT press and hold \<RIT\>
RTS / DTR settings	\<MENU\> \<SET\> \<Connectors\> \<USB SEND/Keying\>
	\<MENU\> \<SET\> \<Function\> \<RTTY Mark Frequency\>
RTTY Mark and Shift	or \<Shift Width\>
RTTY edit messages	RTTY mode \<MENU\> \<DECODE\> \<TX MEM\> \<EDIT\>
RTTY messages via keypad F1 –	\<MENU\> \<SET\> \<Connectors\> \<External Keypad\>
F4	\<RTTY\> \<ON\>
Scan settings	\<MENU\> \<SCAN\>
Screen saver	\<MENU\> \<SET\> \<Display\> \<Screen Saver\>

Screen capture	\<MENU> \<SET> \<Function> \<Screen Capture [POWER] Switch>
Scope settings	Hold \<M.SCOPE> Hold \<EXPD/SET>
SD Card	\<MENU> \<Set> \<SD Card>
Spectrum Scope settings	\<MENU> \<SCOPE> or hold down M.SCOPE then touch and hold EXPD/SET
Speech settings	\<MENU> \<SET> \<Function> \<SPEECH>
Split settings	\<MENU> \<SET> \<Function> \<SPLIT>
Squelch / RF gain control	\<MENU> \<SET> \<Function> \<RF/SQL Control>
Time and date	\<MENU> \<SET> \<Time Set> \<Date/Time>
Tone controls (RX) Treble	\<MENU> \<SET> \<Tone Control/TBW> \<RX> *choose mode* \<RX Treble>
Tone controls (RX) Bass	\<MENU> \<SET> \<Tone Control/TBW> \<RX> *choose mode* \<RX Bass>
Tone controls (TX)	\<MENU> \<SET> \<Tone Control/TBW> \<TX>
Transmit bandwidth	\<MENU> \<SET> \<Tone Control/TBW> \<TX>
Transmit delay	\<MENU> \<SET> \<Function> \<TX Delay>
USB port DTR/RTS settings	\<MENU> \<SET> \<Connectors> \<USB SEND/Keying>
USB SEND Keying	\<MENU> \<SET> \<Connectors> \<USB SEND/Keying>
VFO tuning step	Touch and hold the kHz digits on the VFO
Voice keyer monitor	Voice mode \<MENU> \<VOICE> \<REC/SET> \<SET> \<Auto Monitor>
Voice messages monitor and repeat time	\<MENU> \<VOICE> \<REC/SET> \<SET>
Voice messages external keypad F1 – F4	\<MENU> \<SET> \<Connectors> \<External Keypad> \<VOICE>
VOX settings	Hold \<VOX> button
Waterfall settings	Hold \<M.SCOPE> Hold \<EXPD/SET>
WSJT-X Echo setting	\<MENU> \<SET> \<Connectors> \<CI-V> \<CI-V USB Echo Back> \<ON>